Proceedings in Life Sciences

Coherent Excitations in Biological Systems

Edited by
H. Fröhlich and F. Kremer

With 90 Figures

Springer-Verlag
Berlin Heidelberg New York Tokyo
1983

Prof. Dr. Herbert Fröhlich
Department of Physics, The University of Liverpool, P.O. Box 147,
Liverpool, L69 3BX, U.K.
and
Max-Planck-Institut für Festkörperforschung, Heisenbergstr. 1,
D-7000 Stuttgart 80, F.R.G.

Dr. Friedrich Kremer
Max-Planck-Institut für Festkörperforschung, Heisenbergstr. 1,
D-7000 Stuttgart 80, F.R.G.

Cover: The figure on the cover describes the scheme of a fully decondensed Balbianiring in the field of spherical waves. The diagram is not to scale because the Balbianiring has a diameter of about 50 µm and the spherical waves have a wavelength of a few mm.

ISBN-13:978-3-642-69188-1 e-ISBN-13:978-3-642-69186-7
DOI: 10.1007/978-3-642-69186-7

Softcover reprint of the hardcover 1st edition 1983

2131/3130-543210

Preface

The articles in this volume are based on papers presented at the International Symposium in Bad Neuenahr November 29 - December 1, 1982. The meeting was directed by H. Fröhlich and was sponsored by IBM Deutschland through its Science and Education Programs Department.

Contents

Contributors

You will find the addresses at the beginning of the respective contributions

Coherence in Biology

H. FRÖHLICH

Department of Physics, The University of Liverpool, Oliver Lodge Laboratory, P. O. Box 147, Liverpool, L69 3BX, England, U. K.

The great success of molecular biology arises from the establishment of the atomic structure of biological systems such as DNA or proteins. The activity of these systems does, however, not follow in a simple way from structure as it frequently can be switched on or off. From the point of view of physics this must be expressed in terms of non-linear excitations. Quite different types of excitation often have common general features which has given arise to Haken's synergetics [1]. Establishment and maintenance of such excitations requires the supply of energy. Energy supply, in general, leads to heating. In cases which are of biological interest, however, metabolic energy supply leads to the establishment of organisation, Prigogines's dissipative structures [2]. Whether the one or the other holds must be investigated in detail for each system. No general rule has been found, so far, which would permit a decision between the two possibilities from structure only.

About fifteen years ago it has been conjectured that coherent excitations should play an important role in biological activity; and the conjecture was supported by model calculations ([3], [4], [5]). Such excitations necssarily contain essential non-linear features. In contrast to the realm of linear excitations, systematic investigation then becomes impossible, in principle, as the number of states is too large [6]. Biological systems have, however, developed to fulfil a certain purpose, and it is permissible, therefore, to ask for the purpose of a certain excitation, a question which in physics is permissible only when dealing with machines or similar constructions. Special assumptions must thus be made, and tested against experiment.

The following gives a short survey on the predictions of the theory. The subsequent papers discuss experimental evidence. Coherence determines properties at a space-time point (x^1, t^1) when they are known at another (x,t), such as phase and amplitude of coherent waves. Three basic types of coherent excitations have been proposed:

Coherent Excitations in Biological Systems
Ed. by H. Fröhlich and F. Kremer

A. Coherent excitation of a single polar mode ([5], III C).

B. Excitation of a metastable highly polar state ([5], III B).

C. Vibrations arising from more complex processes and giving rise to limit cycles, or Lotka Volterra oscillations ([5], III E,).

Certain material conditions must be satisfied to permit such excitations. Coherent excitation of a single mode (A) of the band of polar modes may arise from random energy supply at a rate S to all or only some of these modes, provided S exceeds a critical S_o, $S > S_o$, and provided these modes are in strong non-linear, interaction with a heat bath (e.g. cell water) kept at constant temperature. Establishment of the coherent excitation will require a certain time after the start of energy supply. It should be minimal if the energy is supplied directly to the mode that will be excited coherently. The system, thus, possesses storing capability. A range of possible frequencies has been proposed (cf [5], III A), in particular the region of 10^{11} Hz (mm wave region) for sections of membranes, but also lower as well as considerably higher frequencies characteristic for various biological molecules. Furthermore, plasma oscillations of electrons in conduction bands may have to be considered.

A metastable highly polar state (B) may arise from the shape dependence of the electrostatic and elastic energies of a homogenously polarised and deformed system. This possibility should in particular arise in highly polarisable and deformable materials like proteins. This excitation may be supplemented by small conformational changes, such as the transfer of proteins to different positions, e.g. as discussed below by Genzel. It must be realised in this context that the electrostatic energy of a highly polarised molecule may be considerable and can overwhelm changes in local binding energies. The high electric field in a membrane may be expected to lift proteins dissoloved in it into their metastable highly polar state. When in the cytoplasm, on the other hand, the high field due to the polarity of a molecule in the metastable state is likely to be screened by counterions.

The two excitations A and B may have far reaching consequences on the activity of the relevant systems. It may be noted in particular that they can be switched on, or off, depending on the energy supply, giving rise to a particular activation or deactivation. Oscillations of type A yield long range, frequency selective, interactions between systems with equal excitation frequencies; it will be noted that electrostatic interaction is usually screened by counterions. This frequency selective interaction may lead to long range attraction. If this takes place between enzymes and substrates then it proceeds and initiates the usual short range chemical interaction by bringing relevant systems together. We note here a multicausal process, typical for many biological events. It has also been suggested that the selective long range interaction may be relevant for the control of cell division, important in the cancer problem ([5], IV D). Excitation A may also provide facilities for communication within and between cells.

It has been proposed [7] that the high field in membranes would require the use of non-linear optics for the vibrations based on it and hence give rise to self focusing and to involvement in the filamentous microstructure observed in eucaryotic cells.

The highly polar metastable state (B), [8, 9], could represent the active state of enzymes. It stores energy, and, through its internal field can reduce activation energies. This excitation could also be effective in transporting energy from protein to protein. In an array of proteins dissolved in membranes in this way it would act as a large soliton and it has been shown, in fact, that the relevant basic potential energy also leads to ordinary soliton solutions [10].

Considering an array of localised enzymes with activated state described by B, when excitation A attracts substrates, leads to a Lotka-Volterra type of periodic enzyme reaction. Taking account of enzyme-enzyme interaction then yields limit cycles, excitation C. The periodic enzyme excitation - diexcitation in view of the high polarity of B, then causes electric vibrations accompanying the limit cycle, and a sensitivity to weak external fields. This may trigger a collapse of the limit cycle and hence a sudden liberation of considerable energy, as will be reported below by Kaiser.

The consequences of the excitations described above must have decisive influence on the activity of biological systems. They thus might form the dynamic complement to structure. A series of experiments has been designed to establish the existence of these excitations rather than with their biological significance. For excitation of coherent vibrations (A) the most direct method, at first sight, appears to be laser Raman effect when the frequency is above 200 cm^{-1}. For then the ratio of anti- Stokes to Stokes intensity must be noticeabilty larger than in thermal equilibrium. Such experiments have, so far, been carried out on bacterial cells only where a number of experimental difficulties arise. Some of them are connected with the high dilution required to keep the cells in their active state so that a very high degree of synchronisation is necessary (cf [11]). These difficulties have been overcome in the experiment reported by Drissler, below, which indicates strong excitation of a vibration which attracts nutrient molecules into a cell. This interpretation would permit verification by direct measurement of the nutrient influx into a cell which should exceed that arising from diffusion.

Existence of a metastable state (B) in Langmuir-Blodgett layers of proteins is reported below by Hasted. It is reached through application of very high electric fields which indicates implication of the required high dipole moment though at a later stage this will be screened by counterions. Investigations by Mascarenhas [12] with the electret method may be relevant in this context.

Decisive evidence on A arises from experiments on the action of low intensity mm waves on biological materials. It will be noticed that the frequencies are in the range predicted for membrane vibrations. Intensities usually are below $10mW/cm^2$ which excludes thermal effects, as well as direct i.e. non-linear action on an ordinary system [13]. The conclusion must thus be drawn that biological systems have developed an organisation that makes them particulary sensitive to low frequency mm waves. Note that a properly tuned radio is, of course, very sensitive to radio waves of extremely low intensity. This instrument has been specially designed for our needs. There is no evolutionary need, however, for biological systems to be sensitive to external mm waves. The conclusion must then be drawn that the biological systems themselves make use of vibrations in this frequency region i.e. that they use excitation A, or possibly C.

A basic result of the theory requires the energy supplied to cause the excitation to exceed a critical value S_0. This gives rise to a step-like response similar to a phase transition. If the system is in a state close to the step this small external energy supply may trigger off the excitation. Also if the coherent excitation is already excited, but not to its ultimate value, then energy supply at the frequency in question will increase the amplitude. The exact type of excitation may, of course, vary from case to case. Thus apart form the above mentioned case A, a limit cycle of Type C may also be sensitive to weak external fields, as mentioned above.

Sensitivity to low intensity radiation may also arise from cooperative action of a number of charges as is available in the highly polar metastable state B.

The subsequent articles by Grundler, Kremer and Nimtz provide excellent evidence for the existence of such excitations though they can not yet specify details. More direct evidence for the use made of coherence in biological activity requires investigation of specific biological processes. A first step in this direction is discussed below, by Kell. Evidence for attraction between cells, arsing from excitation of membrane vibrations has been obtained for the case of red blood cells by Rowlands, as shown in his article, below. Experiments by Brewer and Bell [14] on the long range interaction between amoebae and anion-exchange particles, and by Roberts et. al. [15] on the intracellular migration of nuclear proteins might also find their explanation in terms of our long range frequency selective forces. Evidence for the existence of relatively low frequency vibrations, with possible significance for cell division is presented below by Pohl. It thus appears that the evidence for the existence of coherent excitations is very strong. So far, however, very little insight has been gained on their specific biological significance. Experiments of a different nature will have to be designed for this purpose. We are dealing with cooperative phenomena, and it would seem that multicellular systems should make use of coherent excitations to a larger extent than single cells. Investigations of differentiated tissues are thus highly desirable.

References

[1] H. Haken, Synergeties, An Introduction 2nd edition, Springer Verlag 1978.

[2] P. Glansdorff and I. Prigogine, Thermodynamic Theory of Structure and Fluctuations. John Wiley and Son, London 1971.

[3] H. Fröhlich, Int. J. Quantum Chem. 2, 641, 1968.

[4] " Theoretical Physics and Biology (M. Marois ed.) p.13 North Holland Press 1969.

[5] " Advances in Electronics and Electronic Physics Academic Press 53, 85, 1980.

[6] " Riv. Nuovo Cimento 3, 490, 1973.

[7] E. Del Giudice, S. Doglia and M. Milani Phys. Lett 90A, 104, 1982.

[8] H. Fröhlich, Nature 228, 1093, 1970.

[9] " J. Collect. Phenom. 1, 101, 1973.

[10] H. Bilz, H. Büttner and H. Fröhlich, Z. Naturforsh 36B, 208, 1981.

[11] H. Fröhlich, Physics as Natural Philosophy (A. Shimony and H. Feshbach ed) MIT Press, Cambridge, 287, 1982.

[12] S. Mascarenhas, Journ. Electrostatics 1, 141, 1975.

[13] H. Fröhlich, Bioelectromagnetics, 3, 45, 1982.

[14] J. E. Brewer and L. G. E. Bell, Exptl. Cell. Res. 61, 397, 1970.

[15] E. M. De Roberts, R. F. Longthorne and J. B. Gurdon, Nature 272, 254, 1978.

P.S.: To avoid unnecessary experimentation: Macroscopically, excitation A will be highly multipolar and hence does not lead to energetically unfavourable emission of radiation.

Coherent Excitations and Raman Effect*

F. DRISSLER and L. SANTO
Max-Planck-Institut für Festkörperforschung, Postfach 800665, D-7000 Stuttgart 80

At first sight measurement of the intensity ratio of antistokes to stokes Raman lines offers a straight forward method for determining the above thermal excitation required by theory. Early measurements /1/, in fact, have confirmed such excitation in the 100 cm^{-1} region. Clearly much more spectacular enhancement should be expected in the higher frequency region near 1000 cm^{-1}, but, so far, stokes spectra only have been taken /2/. Figure 1 shows our result on the time dependence of the E.coli spectra indicating the appearance, and disappearance, of lines as the activity of the bacteria developes. The neighbourhood of the waterband at 1640 cm^{-1} provides the possibility for a rough estimate of intensities. At the high dilution required for the maintainance of the activity of the bacteria, an intensity enhancement of the order 10^5 would be required to find lines of the measured intensity /3/. As a consequence it can be concluded that a number of cells must oscillate coherently, for then the intensity is proportional to the square of the number of oscillating centers. This should explain why, in the time available to us, we were unable to reproduce this result.

Clearly a highly efficient method of synchronisation is required, and we have been advised by Dr. W. Messer (Max-Planck-Institut für molekulare Genetik, Berlin) that the method described in /2/ will only occasionally lead to the required result. To ensure a single cell to be exposed to the dangerous laser light for very brief periods only, we employed a flow arrangement in our experiments. We used the 647.1 nm line of a krypton laser.

---

* In agreement with the authors, this paper was extracted from their notes and supplied with theoretical comments by myself. For personal reasons the authors are unable to continue this research.

H. Fröhlich

Coherent Excitations in Biological Systems
Ed. by H. Fröhlich and F. Kremer

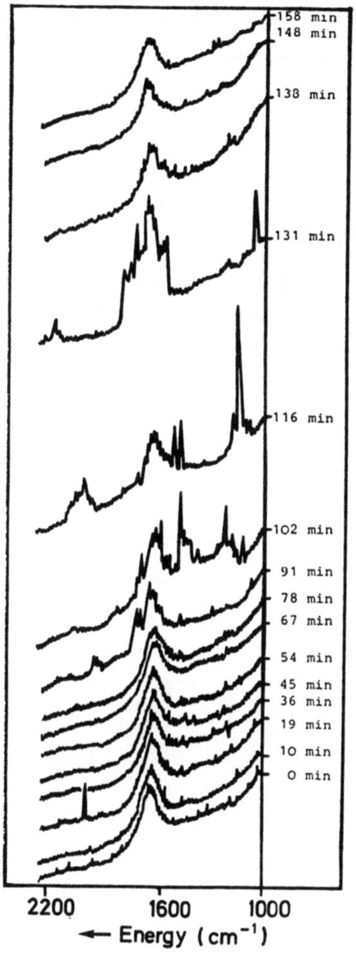

Fig.1.

Time dependent stokes Raman spectra during irradiation with krypton-laser light at 647.1 nm

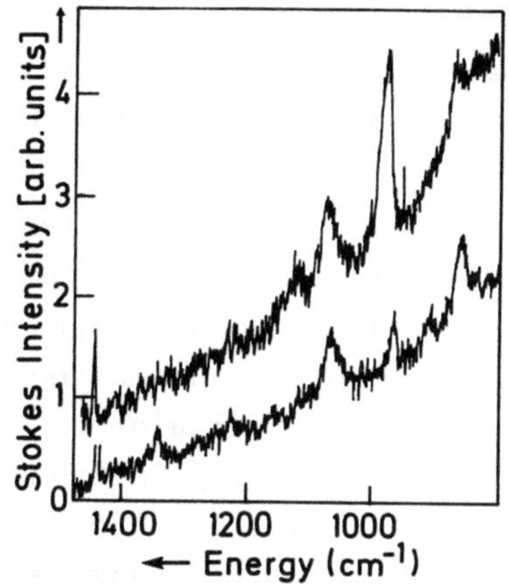

Fig.2.

Stokes Raman spectra after resuspension of E.coli cells in nutrient (upper spectrum) and after 8 hours of cell activity (lower spectrum).

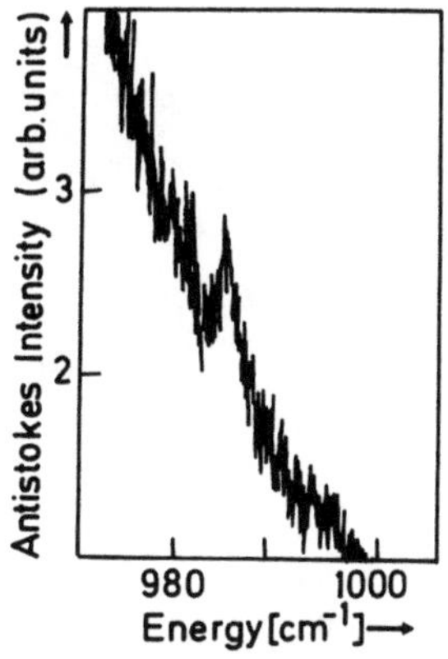

Fig.3. Antistokes Raman intensity from E.coli cells in nutrient. The bands at 980-990 cm^{-1} can be assigned to the $(NH_4)_2SO_4$ and K_2HPO_4 nutrient components.

For the following experiments we used a method of synchronisation suggested by Dr. Messer and described in /4/. Following the thymine starvation described there the cells were resuspended in minimal medium, and we investigated the behaviour of nutrient lines near 980-990 cm^{-1}, which are due to $(NH_4)_2SO_4$ and K_2HPO_4. The results are shown in the subsequent figures; we have not yet shown whether the synchronisation is relevant in this case. Figure 2 shows the stokes line at the beginning and after eight hours, demonstrating the changes of concentrations.

Figure 3 presents the corresponding antistokes line at 980-990 cm^{-1}. Intensities of both stokes and antistokes lines were followed for about 4 hours and are shown in Figure 4a. Assuming the stokes intensity to be in thermal equilibrium, the corresponding antistokes intensity can be calculated. Figure 4b demonstrates that the measured antistokes intensity is higher by a factor between 5 and 15. Figure 4c shows the increase in cell density as measured by the rayleigh scattering of the sample.

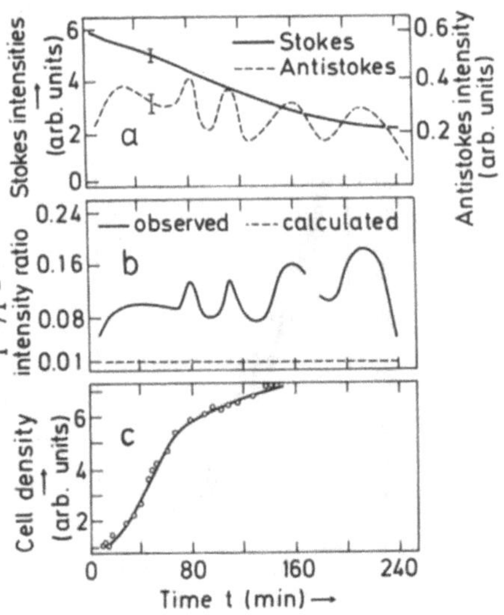

Fig.4a. Time dependence of nutrient stokes and antistokes Raman intensities at 980-990 cm^{-1}.

Fig.4b. Comparison of measured antistokes/stokes intensity with that calculated from the stokes intensity in thermal equilibrium.

Fig.4c. Change of cell density with time.

The only source of energy required here arises from cell metabolism. We can conclude then, that only nutrient molecules in or near a cell are excited, and that their amount of excitation is higher than the above by the very large factor of the ratio of the total number of relevant nutrient molecules to the number of excited ones.

The subsequent attraction is, of course, one of the main consequences of the theory of coherent excitation. A systematic investigation should permit determination of the time and distance a nutrient molecule spends in diffusion; subsequently it will be possible to decide whether nutrient molecules are attracted already outside a cell, or whether the attraction arises within the cell only.

It is regretted that these investigations cannot be continued at present.

REFERENCES

1. Webb, S.J., Stoneham, M.E. and Fröhlich, H. Phys.Lett. A63, 407, (1977).

2. Webb, S.J., Phys.Rep. 60, 201, (1980).

3. Fröhlich, H. in Physics as Natural Philosophy, ed. Shimony, A. and Freshbach, H., MIT Press, Cambridge Masc., p.286, (1982).

4. Nüsslein-Crystalla, V. and Scheefers-Borchel, U., Molec.gen.Genet. 169, 35, (1979).

The Non-thermal Effect on Millimeter Wave Radiation on the Puffing of Giant Chromosomes

F. KREMER[1], C. KOSCHNITZKE[2], L. SANTO[1], P. QUICK[2], and A. POGLITSCH[1]

[1] Max-Planck-Institut für Festkörperforschung, Heisenbergstr. 1, D-7000 Stuttgart 80

[2] Institut für Allgemeine Genetik; Universität Hohenheim, Garbenstr. 30, D-7000 Stuttgart 70

ABSTRACT

A non-thermal influence of millimeter wave radiation (swept in frequency from 64.1 GHz to 69.1 GHz, sweeptime 6 s, and with stabilized frequencies of 67.200 ± 0.001 GHz and 68.200 ± 0.001 GHz, power density $\leqslant$ 5 mW/cm^2) on the puffing of giant chromosomes of the midge *Acricotopus lucidus* (Diptera, Chironomidae) was found. The effect is manifested as a reduction in size of a specific puff that expresses genes for a secretory protein. The experiments were carried out blind and the effect could be established to a level of significance of $P < 0.5$ %. Concerning the very low photon energy of mm-waves compared to the thermal energy kT, we conclude that the coherence of the radiation must be decisive for the observed effect. Our result could possibly be understood by H. Fröhlich's (1-2) theory of coherent electric vibrations in biological systems.

INTRODUCTION

The non-thermal influence of millimeter wave radiation on biological systems is a topic of considerable importance not only for the understanding of the mechanisms of interaction between electromagnetic radiation and living systems (1-2) but also for the establishment of safety standards against possible hazards from microwaves. Several biological effects of low-intensity millimeter wave radiation have been reported (3-7), among them genetic effects in Drosophila melanogaster (4,6). In studying the puffing pattern of giant chromosomes in salivary glands of the midge *Acricotopus lucidus* we found that a certain puff, the Balbianiring BR2 in the chromosome II exhibits reductions after irradiation with millimeter waves (Fig.1).

Coherent Excitations in Biological Systems
Ed. by H. Fröhlich and F. Kremer

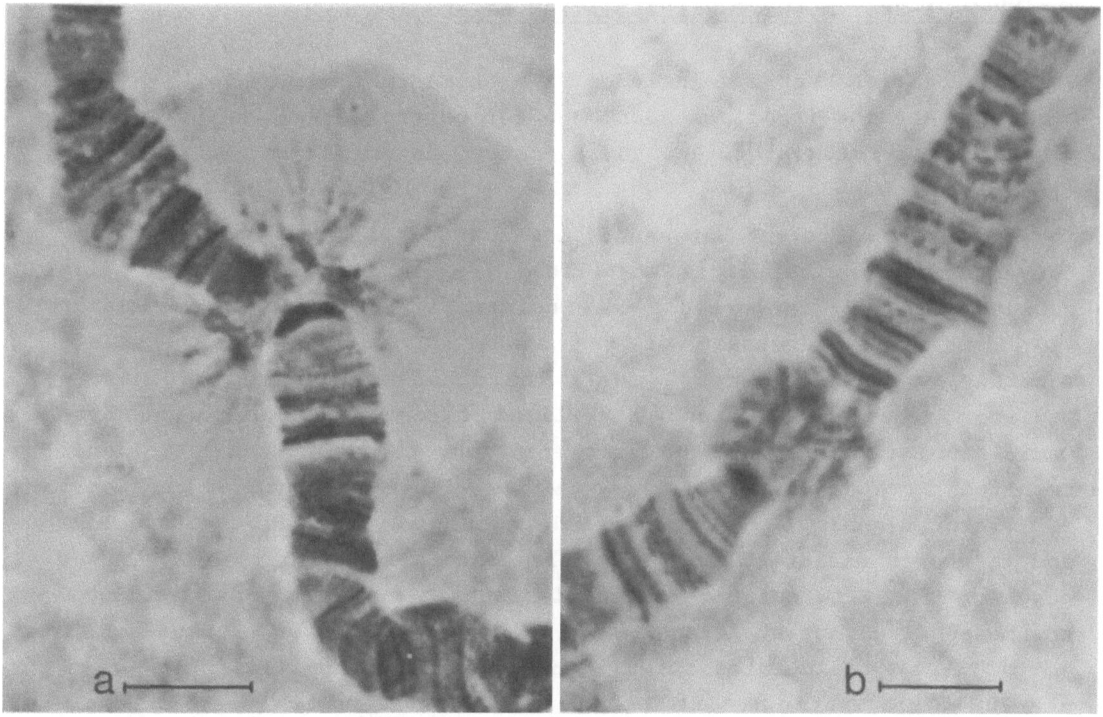

Fig.1. a) Balbianiring BR2 (control), fully decondensed.
b) BR2 locus after (2 hrs) irradiation with millimeter waves. The BR2 has regressed. The chromatine fibrills are totally condensed and the surrounding puff-material (ribonucleoprotein) has disappeared. Scale bars: 10µm.

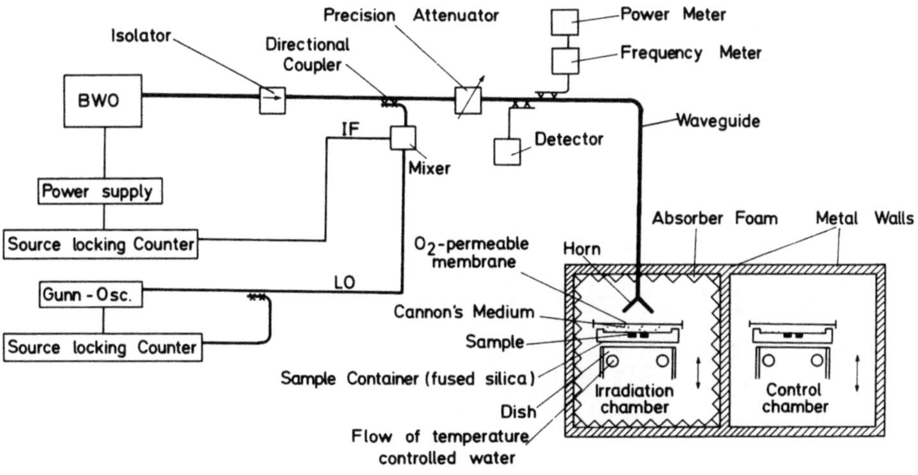

Fig.2. Schematic description of the mm-wave system.

THE MM-WAVE-SYSTEM AND THE IRRADIATION SET-UP

The irradiation set-up included two electromagnetically isolated chambers; one for irradiation and one for control. As mm-wave source a Backward-Wave-Oscillator (RWO 80, Siemens) was used (Fig.2)

The frequency was swept between 64.1 GHz and 69.1 GHz (Sweeptime: 6s). For experiments with single stabilized frequencies two source-locking counters (EIP 471) were employed. One stabilized the local oscillator (10 GHz - 15 GHz, accuracy of the stabilization: ± 2 Hz, absolute accuracy of the counter: 10^3 Hz). The other one counted the intermediate frequency (IF) and stabilized directly the Backward-Wave-Oscillator, (accuracy of the stabilization: ± 1 MHz). This IF signal resulted from the mixing of the mm-waves with a harmonic of the local oscillator signal (LO). The forward power was measured as 20 ± 2 mW and the power reflected by the sample container as 2 ± 0.5 mW. Thus a power of 18 ± 2 mW entered the sample container. After passing through the medium layer of at least 200 µm thickness the power was reduced to 4.5 ± 0.5 mW (calculated by using the absorption coefficient $\alpha = 70\ cm^{-1}$ of water (8) at 7×10^{10} Hz for the medium). This resulted in a power density of less than 5 mW/cm^2 (horn area 1.6 cm^2).

The salivary gland was placed in a sample container which consisted of a fused silica plate with an indented circle of radius 40 mm and depth 0.2 ± 0.02 mm (Fig.3). In its centre two further circular incisions of diameter 2 mm and depth 0.3 ± 0.02 mm were prepared. One salivary gland was put in each of these last indentations. Cannon's medium (9) was added and then the sample container was covered by an oxygen-permeable membrane. The glands are sac-like and have a diameter of about 0.25 - 0.30 mm. Thus the layer of aqueous medium between the

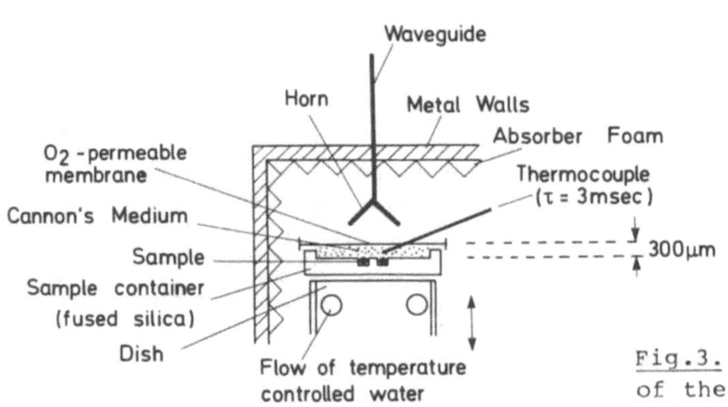

Fig.3. Schematic description of the irradiation set-up.

membrane and the glands was at least 0.2 mm thick. The sample container was positioned on a fused silica temperature controlled dish (9.0 ±0.1°C). To average over the standing wave pattern between the horn antenna and the sample container the whole dish was moved up and down in front of the horn with an excursion of several wavelengths (as indicated by the arrow in Fig.3).An identical dish was mounted in the control chamber.

To measure the microwave induced temperature increase of the sample a micro-miniature thermal probe (diameter: 0.2 mm, thermal rise-time: 3 ms) was used (Fig.3). By inserting this with an x-y drive quickly into the sample (time needed: about 5s) its temperature was measured in less than a second. This time is short compared to the thermal rise-time of the whole sample (about 40 s), so that the disturbance of the near-field of the horn due to the metallic thermo-couple could only negligibly influence the sample temperature. By measuring the sample-temperature without and with radiation present the microwave-induced temperature increase was found to be less than 0.3°C for a forward power of 20 mW. For the actual irradiation experiments the thermal probe was removed.

BIOLOGICAL SYSTEM

We used larvae of the midge *Acricotopus lucidus* in their fourth larval instar. The larvae were dissected and their paired salivary glands taken out for the experiment. The salivary glands are composed of two clearly different cell-types. These cell types correspond to morphologically distinct lobes, designated main lobe and anterior lobe. Each main lobe consists of about 50-60 cells whereas the smaller anterior lobe has 12-20 cells. The nuclei of all cells contain 3 polytene giant chromosomes. In these chromosomes cell type specific Balbianirings are developed: Balbianiring BR1 and BR2 on chromosome I and II in the main lobe, Balbianiring BR3 and BR4 on chromosome I and II in the anterior lobe. The appearance of these Balbianirings can be directly inspected with cytological methods, moreover they are reflected by the production of cell-type specific secretion proteins at the cytoplasmic level (9-14). Thus this system permits the simultaneous examination of different Balbianiring-patterns in 2 diversely differentiated cell-types. Since in preceeding irradiation experiments no obvious reactions at other Balbianiring sites had been observed, we only examined the BR2.

The puffing-phenomenon is a highly complex process which does not depend solely on transcriptional activity, i.e., enzymatic activity of RNA-polymerases. It comprises interactions between transcription, RNA-processing, packaging of RNA with distinct nuclear proteins and storage or transport of ribonucleoprotein-products. Especially for the appearance of the BR2, post-transcriptional processes are of particular importance. This is obvious when one considers some of the unusual properties of this predominant, cell-specific puff. First, its large size does not correspond to its relatively low rate of transcription as compared to all other so far examined puffs of the Balbianiring type (15,16). Secondly, this extraordinary high ratio of nonhistone proteins to RNA at the BR2-puffing border points to a leading role for the post-transcriptional events in the regulation of the BR2-gene expression (16). When judging the cellular consequences of BR2-regression it is of interest that a cytological disappearance of the BR2-puff only has been demonstrated to be correlated with an exclusive inhibition of RNA-synthesis at the BR2-locus and loss of synthetic capacity for a specific secretion component (11-14) of the saliva.

EXPERIMENTAL PROCEDURE

For the experiment the paired salivary glands from larvae of the fourth larval instar of *Acricotopus lucidus* were dissected. One gland was placed in the irradiation chamber and the other in the control chamber. Immediately after irradiation (2 hrs) each gland was fixed with ethanol-acetic acid (3:1). The samples were stained for squash preparations. Three different types of Balbianirings BR2 were distinguished: normal, weakly reduced, and strongly reduced. Correspondingly, for one gland the numbers of normal (m_1) and weakly reduced (m_2), and strongly reduced (m_3) Balbianirings BR2 were determined and the reduction probability $r_3 = \frac{m_3}{m_1 + m_2 + m_3}$ was calculated for the sample in the (sham)-irradiation chamber ($r_3^{irr.}$) and the control chamber ($r_3^{contr.}$).

All experiments were carried out <u>blind</u>, i.e, the examining biologist did not know which sample was irradiated and which had served as control. Furthermore, without informing the biologist three different types of experiments were carried out: I. sham-exposure experiments, II. experiments, in which the sham-exposed sample was warmed up by 2.5°C over the control by extra heating, III. irradiation experiments.

RESULTS AND STATISTICAL ANALYSIS

For the statistical analysis the difference Δr_3 between the probability of a strong reduction r_3 of a gland placed in the irradiation chamber $r_3^{irr.}$ and in the control chamber $r_3^{contr.}$ is calculated (s. Tab.1 and Fig.4). Two statistical tests were employed: The ±-Test and the U-Test of Mann-Whitney (17). The ±-Test calculates the probability to find k or more "successes" ($\Delta r > 0$) in n trials assuming a binomial distribution with $p = 1/2$ (i.e. no difference between irradiation and control). The U-Test measures the probability that two experimental series belong to the same distribution.

Type		n	$\overline{\Delta r_3}$	s	k	l	P_1	P_2
I	sham-exposed	43 (2363)	0.0085	0.012	23.5	19.5	27.1	/
II	sham-exposed with additional heating of 2.5°C	20 (1181)	0.0060	0.011	10.5	9.5	41.2	27.4
IIIa	64.1 GHz - 69.1 GHz Power density: < 5 mW/cm^2	35 (2000)	0.0523	0.020	22.5	12.5	4.5	4.2
IIIb	67.200 ± 0.001 GHz Power density: < 5 mW/cm^2	13 (707)	0.0606	0.017	11.5	1.5	0.2	0.5
IIIc	68.200 ± 0.001 GHz Power density: < 5 mW/cm^2	17 (931)	0.0525	0.012	14.5	2.5	0.1	0.4

Tab.1. n: number of gland pairs (number of chromosomes in brackets) $\overline{\Delta r_3}$: mean value of Δr_3 for the different types of experiments. s: standard deviation of the mean value. k: number of "successes" with $\Delta r_3 > 0$. l: number of "failures" with $\Delta r_3 < 0$; in cases of $\Delta r_3 = 0$ we counted $k = l = 0.5$.
P_1: probability in percent for k or more successes in n trials assuming a binomial distribution with $p = 1/2$.
P_2: probability in percent that the experimental series (II, IIIa-IIIc) belong to the same distribution as sham exposed series (I) according to the U-Test of Mann-Whitney (17).

In Fig.4 the values corresponding to Tab.1 are plotted.

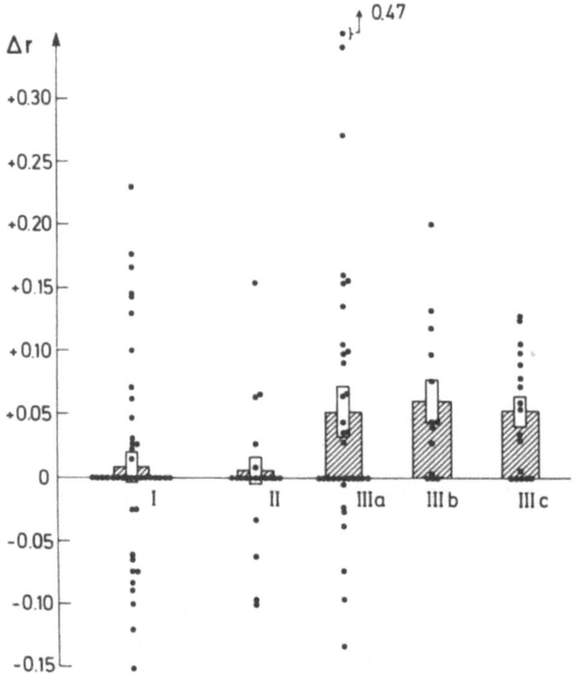

Fig.4. The dots represent the difference $\Delta r_3 = r_3^{irr.} - r_3^{contr.}$ for one gland pair each in the sham exposure (I), the sham exposure with additional heating of 2.5°C (II) and the irradiation experiments (III). The shaded columns indicate the mean value of all Δr in one type of experiment, the empty columns the standard deviation of the mean value.

In the previous analysis only the strong reductions m_3 were regarded. Also a weighted reduction probability defined as

$$r_w = \frac{1 \cdot m_1 + 2 \cdot m_2 + 3 \cdot m_3}{m_1 + m_2 + m_3}$$

was calculated for the gland in the (sham)-irradiation chamber ($r_w^{irr.}$) and the control chamber ($r_w^{contr.}$). Then the difference $\Delta r_w = r_w^{irr.} - r_w^{contr.}$ was determined. The result is shown in Tab.2.

Type		n	$\overline{\Delta r_w}$	s	k	l	P_1	P_2
I	sham-exposed	43 (2363)	0.0023	0.0418	24	19	27.1	/
II	sham-exposed with additional heating of 2.5°C	20 (1181)	0.0202	0.0520	10	10	58.8	37.8
IIIa	64.1 GHz - 69.1 GHz Power density: < 5 mW/cm^2	35 (2000)	0.1623	0.0438	27	8	0.09	1.0
IIIb	67.200 ± 0.001 GHz Power density: < 5 mW/cm^2	13 (707)	0.1613	0.0436	11	2	1.1	2.0
IIIc	68.200 ± 0.001 GHz Power density: < 5 mW/cm^2	17 (931)	0.1766	0.0183	17	0	0.001	0.1

Tab.2. n: number of gland pairs (number of chromosomes in brackets) $\overline{\Delta r_w}$: mean value of Δr_w for the different types of experiments. s: standard deviation of the mean value. k: number of "successes" with $\Delta r_w > 0$. l: number of "failures" with $\Delta r_w < 0$; in cases of $\Delta r_w = 0$ we counted $k = l = 0.5$.
P_1: probability in percent for k or more successes in n trials assuming a binomial distribution with $p = 1/2$.
P_2: probability in percent that the experimental series (II, IIIa-IIIc) belong to the same distribution as sham exposed series (I) according to the U-Test of Mann-Whitney (17).

CONCLUSIONS

1. While for both sham-exposed series (I, II) no significant effect is observed, a highly significant effect is found when the mm-wave radiation is present (III) (Tab.1 and 2). Note that the mm-waves stabilized in frequency (IIIb and IIIc) seem to be more effective than swept frequencies (IIIa).

2. Since in the experiment of type II the sham-exposed sample was warmed up by more than the eight-fold microwave induced temperature increase, the failure to find a significant effect proves that the influence of the irradiation cannot be of thermal origin. It should be noted that a localized overheating of the sample above the temperature of the surrounding medium is not possible because of the facts that the sample itself contains about 90 % water and that the mm-wave absorption of the protein-material is about hundredfold smaller than that of water (18,19). So the thermal properties of the sample can be considered as that of water while its mm-wave absorption cannot exceed that of the surrounding medium.

3. The energy of a mm-wave photon $h\nu$ (for $\nu = 7 \times 10^{10}$ Hz: 2.9×10^{-4} eV) corresponds to less than 1/200 of the thermal energy kT at 300 K. Therefore, a single photon process cannot explain the influence of the irradiation. So we must conclude that the coherence of the applied radiation is decisive for the observed effect.

4. The observed effects occur at a power density of less than 5 mW/cm^2 This corresponds to a field strength in the medium of about 0.7 V/cm. This value is much smaller than physiological field strengths (e.g. field strength across the nerve membrane: 10^5 V/cm).

5. Assuming a dipole interaction of the outer electromagnetic field $\vec{E}$ with the biological system the interaction energy $\vec{\mu} \cdot \vec{E}$ has to be large compared to the thermal energy kT of the system:

$$\vec{\mu} \cdot \vec{E} \gg kT$$

Inserting for the field strength $\vec{E}$ 0.7 V/cm a value of $|\vec{\mu}| > 10^8$ Debye results for the dipole moment (4.8 Debye is the dipole moment of a dipole with one elementary charge and a length of 1 Å). For a dipole of length l and with N elementary charges e it holds:

$$\mu = N \cdot e \cdot l$$

From that one finds

$$N \cdot l \quad > 10^7 \text{ Å} = 1 \text{ mm}$$

an unrealistic value. This shows that the observed effects cannot easily be understood in terms of a simple dipole interaction. An interaction of higher multipoles with the outer electromagnetic field is also not probable.

6. The observed effects could be understood following Fröhlich's (1-2) conjecture of coherent electric vibrations in biological systems. Their frequency - if based on membrane oscillations - is estimated to be of the order of 10^{10} Hz to 10^{11} Hz. So the observed effect could possibly be comprehended by assuming that the externally applied radiation field influences the excitations of the biological system.

7. Our result could be of importance in the discussion of safety standards with regard to possible hazards from millimeter wave radiation. It is shown that millimeter waves of a power density less than 6 mW/cm^2 exert a non-thermal influence on the chromosomes of our eucaryotic system. This level is below the safety standard of 10 mW/cm^2 in most European countries and the USA.

References

1) Fröhlich, H. Intern.J.Quant.Chem. 2, 641-649 (1968).
2) Fröhlich, H. in: Advances in Electronics and Electron Physics, Academic Press 53, 85-152 (1980).
3) Webb, S.J. and Booth, A.D. Nature 222, 1199-1200 (1969).
4) Devyatkov, N.D., Bazanova, E.B., Bryukhova, A.K, Vilenskaya, R.L., Gel'vich, E.A., Golant M.B., Landau,N.S., Mel'nikova, V.M., Mikaélyan, N.P., Okhokhonina, G.M., Sevast'yanova, L.A., Smolyanskaya, A.Z., Sycheva, N.A., Konrat'eva, V.F., Chistyakova, E.N., Shmakova, I.R., Ivanova, N.B., Treskunov, A.A., Manoilov, S.E., Strelkova, M.A., Zalyubovskaya, N.P., Kiselev, R.I., Gaiduk, V.I., Khurgin, Y.I. and Kudryashova V.A., Scientific sessions of the division of general physics and astronomy (Jan. 17-18, 1973), USSR Academy of Sciences, Sov.Phys.-Usp. (Translation) 16(4), 568-579 (1974).

5) Berteaud, A.J., Dardalhon, M., Rebeyrotte, N. and Averbeck, D., C.R.Acad.Sc.Paris, 281, 843-846 (1975).
6) Dardalhon, M., Berteaud, A.J. and Averbeck, D. in: Int.Symp. on Biological Effects (URSI), Airlie (USA) C 2, 25 (1977).
7) Grundler, W. and Keilmann, F., Z.Naturforsch. 33c, 15-22 (1978).
8) Szwarnowski, S. and Sheppard, R.J., J.Phys.E 10, 1163-1167 (1977).
9) Ringborg, U., Daneholt, B., Edström, J.E. Egyhazhi, E. and Rydlander, L.J., Molec.Biol.51, 679-686 (1970).
10) Mechelke, F. Chromosoma 5, 511-543 (1953).
11) Panitz, R. Biol.Zbl. 86, Suppl., 147-156 (1967).
12) Baudisch, W. and Panitz, R., Exp.Cell Res. 49, 470-476 (1968).
13) Panitz, R. in: Results and Problems in Cell Differentiation, Vol.IV, (ed.W.Beermann) 209-227 (Springer Verlag Berlin-Heidelberg-New York, 1972).
14) Baudisch, W. in: Biochemical Differentiation in Insect Glands (ed.W.Beermann) 197-212 (Springer Verlag Berlin-Heidelberg-New York, 1977).
15) R. Panitz, E. Serfling, V. Wobus, Biol. Zbl. 91, 359 (1972).
16) P. Quick, Doctoral thesis, University of Hohenheim, 1982.
17) Siegel, S. Non-parametric Statistics for the Behavioural Sciences (McGraw-Hill, New York, 1956).
18) F. Kremer, L. Genzel, Invited talk on the 6th International Conference on Infrared and Millimeter Waves, paper T-1-4, Miami Beach, December 1981.
19) L. Genzel, F. Kremer, A. Poglitsch, G. Bechtold accepted for publication in Biopolymers.

Nonthermal Resonant Effects of 42 GHz Microwaves on the Growth of Yeast Cultures

WERNER GRUNDLER[1], FRITZ KEILMANN[2], VERA PUTTERLIK[1], LANNIANTI SANTO[2], DIETRICH STRUBE[1], and INGRID ZIMMERMANN[1]

[1] Gesellschaft für Strahlen- und Umweltforschung, D-8042 Neuherberg
[2] Max-Planck-Institut für Festkörperforschung, D-7000 Stuttgart 80

ABSTRACT

Repetition of our earlier experiment has confirmed that the growth rate of aqueous yeast cultures is affected by weak microwave radiation in a frequency-selective manner. The extensions of the experimental procedure included a refined frequency stabilization, refined power recording, impedance matching elements, two geometrically different antenna structures and two recording photometers. Laser thermometry was employed to locate any hot spots.

The result is that depending on the frequency (near 42 GHz) both increases and decreases of the growth rate occur, within resonance bands of only 8 MHz full width at half maximum. The effects are reproducible over long periods (years) and with different irradiation geometries. A rather flat intensity dependence is found, e.g. constancy of the effect at 41782 MHz for nearly an order of magnitude variation of applied power.

Microwave-induced heating was measured to amount to 0.6°C for the highest power applied, and to be spatially homogeneous within 0.02°C throughout the stirred liquid culture. This calls for a nonthermal origin of the effect. On the other hand single-quantum effects seem unlikely since the microwave photon energy is much smaller than the thermal energy kT. These findings therefore support theoretical models which suggest the existence of specialized many-quanta receivers, e.g. coherent molecular oscillations.

Coherent Excitations in Biological Systems
Ed. by H. Fröhlich and F. Kremer

INTRODUCTION

Theoretical considerations by Fröhlich have suggested that living objects might contain systems exhibiting large-amplitude oscillations at high frequencies /1-3/, a conjecture with quite far-reaching consequences for our understanding of biological processes. Excitation of such systems is thought to ordinarily occur through metabolic pumping. Radiative processes on the other hand would open such systems to spectroscopic investigations. A trigger action of resonantly applied external radiation has been discussed in a detailed nonlinear oscillator model ("limit cycle") by Kaiser /4/.

In this respect, our earlier experiment /5/ on the influence of low-intensity millimeter microwaves on the growth of yeast was very interesting. It did show a resonant ($Q \approx 10^3$) action on the growth in contrast to all spectroscopic evidence /6/ which supports the usual expectation that only broadband, nonresonant absorption features exist in biological materials at these frequencies. Since the pioneering work of Webb on E. coli /7/, further evidence of resonant action of millimeter waves has been established in a number of systems /8-11/. Other studies report nonthermal effects on further biological systems, investigated at certain fixed millimeter wave frequencies /12-20/. On the other hand some attempts have failed to demonstrate effects /21-25/. Recent results of USSR groups are described in /33/.

The purpose of this contribution is to sum up our recent activity /26-32/ concerning the nonthermal resonant influence of microwaves on the growth of aqueous yeast cultures.

EXPERIMENT

As in ref. /5/, we used a diploid, homozygot and isogene wild type strain of Saccharomyces cerevisiae (type 211). The cells were grown on agar plates for three days at 30°C. These plates were stored at 4°C. Cells for liquid suspension were taken from these plates after 10 to 16 days. With starting concentrations of about 3.10^5 cells/cm^3 these cultures were held in small glass cuvettes equipped both with mechanical stirrers and with submersible teflon antennae used for coupling in the microwaves.

The simultaneous growth of two yeast cultures was measured in two double-beam spectrometers (Beckman Acta CIII and Beckman M24), set at 550 nm. The reference cuvettes contained plain growth medium. The extinction or optical density E due to the yeast cells is dominated by light scattering /26/. Since a certain part of the scattered light arrives at the detectors the photometers measure a somewhat reduced extinction. Because of the different optical systems employed in both instruments, the calibration curves (Fig.1) vary.

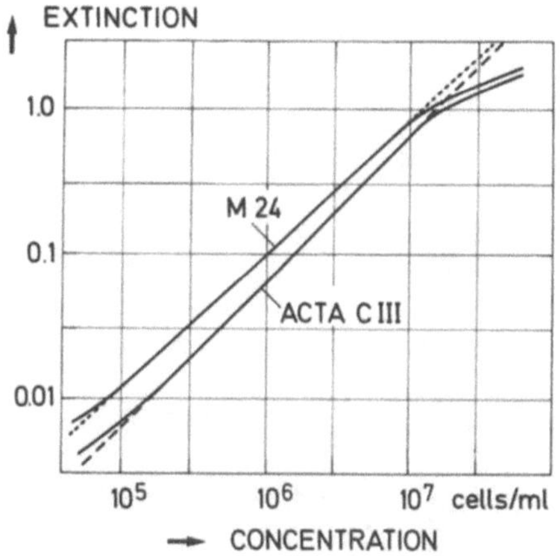

Fig.1. Calibration of the two photometers used. First the concentration of a stationary phase yeast culture was obtained by microscopic counting, and then the extinction was measured for a series of dilutions (in medium), in a 1 cm pathlength cuvette.

The apparent optical density signals V_{OD} were amplified in logarithmic amplifiers and continuously recorded as ln V_{OD} vs. time t. After several hours of lag-phase the yeast cultures attain the exponential growth phase which manifests itself in a linear trace of which the slope is read out to give the growth rate $\mu = t^{-1} \ln V_{OD}$. A correction factor of 1.07 is applied to obtain μ in the M24 to correct the slightly nonlinear response of this instrument (Fig.1).

The growth rate μ is found to be about 0.55 h^{-1} at 31°C, with only weak dependence on temperature $d\mu/\mu dT$ = 0.027 or 0.019°C^{-1} at 31 or 33°C, respectively (cf. Fig.3 in /28/). Both sample cuvettes were stabilized in temperature from a common water thermostat, at 30.7°C in all experiments reported here.

The sample cuvettes were equipped with two types of submersible antennae. One was the "fork"-shaped antenna already used in the earlier experiments (Fig.5 in /5/). The other was of a much simpler cylindrically symmetric shape ("tube" Fig.2)

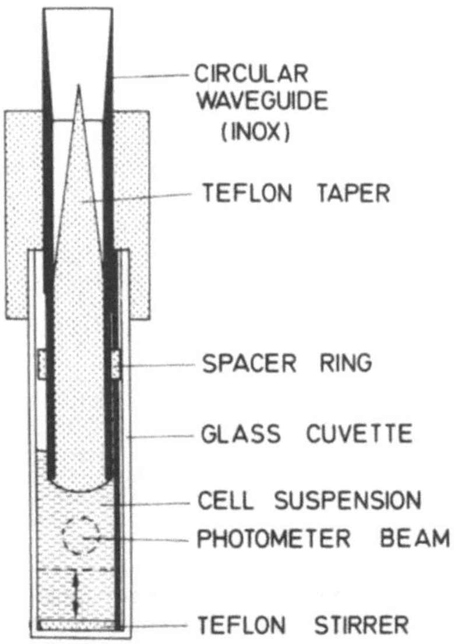

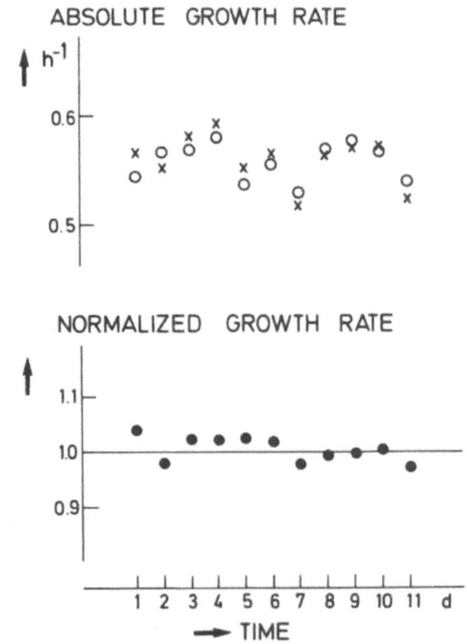

Fig.2. Scale drawing of a 1 cm pathlength photometer cuvette equipped with mechanical stirrer and "tube" shaped teflon antenna.

Fig.3. Growth rates obtained in consecutive experiments at 30.7°C with photometer Acta CIII (x) and M24 (o) using fork antennae (no irradiation). The lower plot gives the ratio $\bar{\mu} = \mu_x/\mu_o$.

In spite of a rather strict preparation procedure to obtain standardized cultures the growth rate μ is found to vary from day to day, as seen in Fig.3, within a range of ± 10 %.

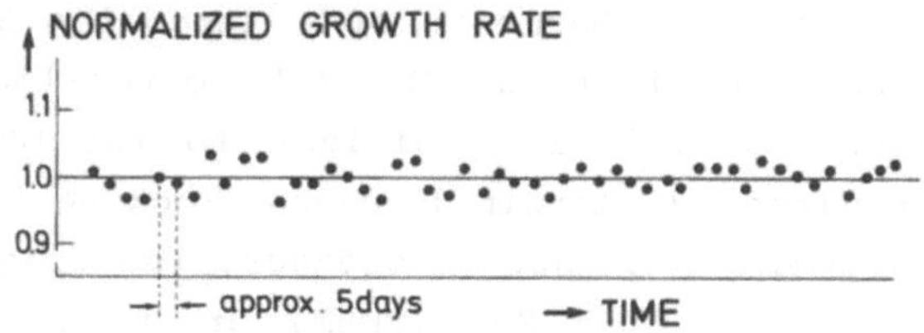

Fig.4. Normalized growth rates $\bar{\mu}$ with no irradiation applied, obtained in a long series (see also Fig. 9b), using tube antennae.

These data show that the normalized growth rate observed for simultaneously growing cultures stays within rather narrow fluctuation limits of 1 ± 0.04. Therefore we carried out the irradiation experiments by applying microwaves to the sample in the photometer Acta CIII only, and by plotting normalized growth rates μ_x/μ_o (in Fig.9 below). Usually the power was turned on after one hour of exponential growth and μ_x was determined from the slope thereafter (Fig.8).

The microwave system is shown schematically in Fig.5 below:

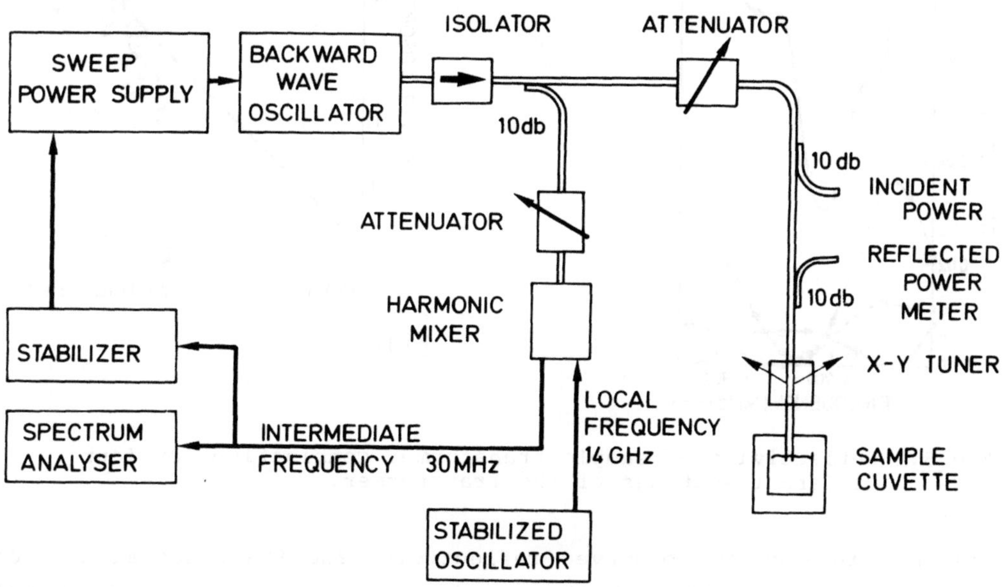

As a microwave source we used a Siemens BWO 60 backward wave tube capable of emitting up to 50 mW in the range 40 to 60 GHz. This oscillator was controlled by a Micro-Now 702/703 C power supply connected to an Alfred 650 sweep unit. The frequency was stabilized in respect to the third harmonic of a Marconi 6150 oscillator which in turn was phase-locked to an EIP 371 source-locking counter. Feedback from a Micro-Now 211-Cl stabilizer to the G2 grid of the BWO power supply resulted in marginal phase locking only, with a retain bandwidth exceeding ± 5 MHz. Altogether, the frequency was stabilized to a bandwidth of ± 0.3 MHz (FWHM) around a set point which was absolutely calibrated to better than 0.1 MHz.

The microwave line to the sample consisted of standard components with an overall length of 1 m. Forward and backward power were monitored with a Hewlett-Packard 432A power meter and calibrated diodes. The x-y tuner was adjusted in each experiment to minimize the reflected power. Standing wave resonances of this line were measured with the filled sample cuvette in place (fork antenna).

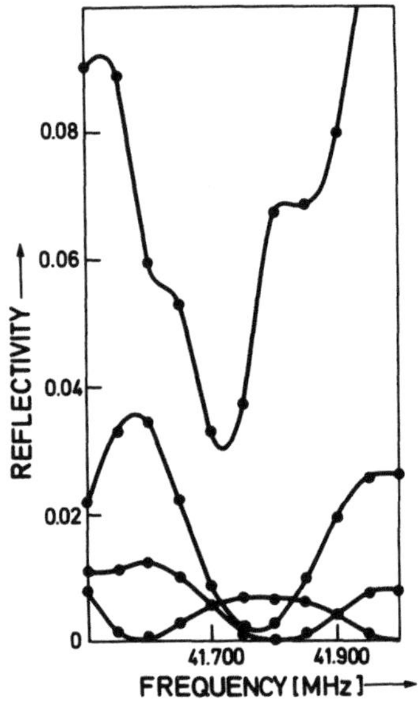

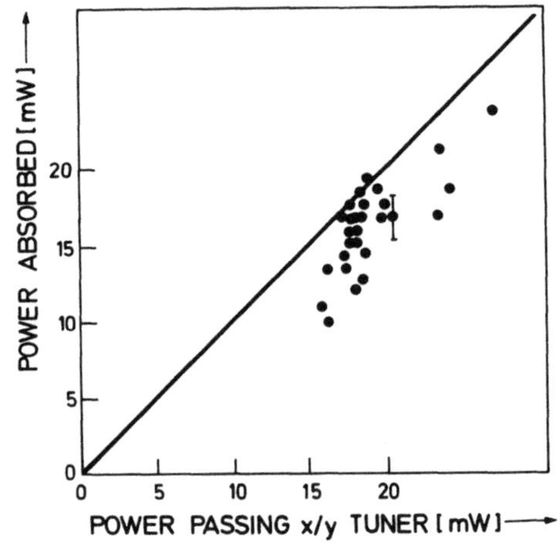

Fig.7.

Fig.6. Reflectivity of sample/transformer combination for four different settings of the transformer.

The magnitude of the observed reflectivity and the spectral periodicity (roughly 250 MHz) of the standing wave resonances can be well understood from simple estimations involving the length as well as the reflection coefficients at the ends of the transmission line /27,30/. In the actual irradiation experiments the frequency was first stabilized and the transformer then adjusted for zero reflectivity. The forward power passing the transformer was then continuously recorded.

In addition a direct recording of the power absorbed in the sample cuvette was routinely performed. Quite similarly as in our earlier experiment /5/ we measured the slight steady-state temperature rise of the sample which occurred due to the absorption of microwaves /29,31/. A calibration as described in /5/ gave a calorimetric response of 0.016°C per mW absorbed power. The price to pay for this rather direct power measurement is that the irradiated sample's temperature was not fixed at 30.7°C but had an over-temperature between 0.16°C to 0.4°C for the usually applied power range from 10 to 25 mW.

Figure 7 demonstrates the correlation obtained with both independent power measurements, for an experimental series where the frequency was varied between 41630 and 41800 MHz. The straight curve is expected for the ideal case of no losses (between the forward directional coupler and the sample cuvette) and of no calibration errors. The data points show that 90 ± 10 % of the applied power is absorbed in the sample, the fluctuation being probably due to a loss occurring at the demountable waveguide joint between the transformer and the sample cuvette. In the remainder of this paper we use the absorbed power to measure the irradiation intensity.

On several occasions following our publication /5/ it has been critisized that standing waves within the fork-shaped teflon antenna could lead to hot spots and apparently resonant effects. Of course we do not claim a constant irradiation intensity along the teflon/water surface, but we rather expect a certain partially standing wave pattern /27,30/. Local intensities up to three times higher or lower than the average are estimated from the known reflectivity of about 0.5 at the teflon/-water interface. This pattern should, however, not vary noticeably on an altogether 0.5 % tuning of the frequency as in our experiment, the reason being that the teflon structure is only a few wavelengths in size. In spite of this we conducted a laser interferometric experiment /30/ to visualize the microwave induced heating pattern inside the water just next to the radiating teflon surface. This method is based on the fact that the refractive index of water changes by 10^{-4}/°C near 30°C (at the wavelength 633 nm of the HeNe laser). The presence of light-scattering particles, e.g. yeast cells in the suspension, and the rapid stirring of such a suspension does not hinder the optical phase measurement, which we obtained from photographing Mach-Zehnder interference patterns. The viewing direction was coincident with the photometer beam's direction (Fig.5 in /5/). With 1 cm pathlength in the water, one full interference fringe corresponds to a temperature increase, averaged along the line of sight, of 0.63°C. Several interference fringes could be observed when the stirrer was not in action and when an especially high-power klystron was employed (42 GHz, absorbed power 300 mW). An analysis of the experiments showed that sizable irradiation indeed occurred on any part of the interface, with no prominent hot spots visible. Furthermore with the usual conditions (stirrer on and steady-state irradiation up to 40 mW (absorbed) power) the temperature is homogeneous throughout the sample suspension within 0.02°C, i.e. the latter is the maximum possible temperature difference between any points in the suspension, including the boundary layer near the teflon surface.

Furthermore in this context it should be noted that in our experiment the possibility of any significant microscopic hot spots on cellular and subcellular scales can be rigorously ruled out /27,30/, from heat conduction arguments. For example, microwave-induced steady-state overtemperatures are estimated to be much smaller than 10^{-5} °C for any part of the yeast cells.

EXPERIMENTAL RESULTS

A total of 331 growth experiments were carried out. In 28 cases the results had to be discarded because of bacterial infection or some failure in the irradiation or recording electronics. Fork or tube antennae were alternatively employed for series of different time periods. The resulting normalized growth rates for those runs where no irradiation was applied are shown in Figs. 3 and 4. This demonstrates that the growth rate of two simultaneously growing yeast cultures can be reliably measured in the two photometers. A maximum deviation of 4 % from the mean has been observed. Rather conservatively, we assign for the following display of the results of the irradiation experiments an error bar of ± 4 % to the normalized growth rate of each irradiation experiment. When the irradiation was turned on the occurence of a sizable effect was regularly indicated by a kink in the plotted growth curve, as shown in Fig.8.

The normalized growth rates of the irradiation experiments are plotted vs. irradiation frequency in Fig.9.

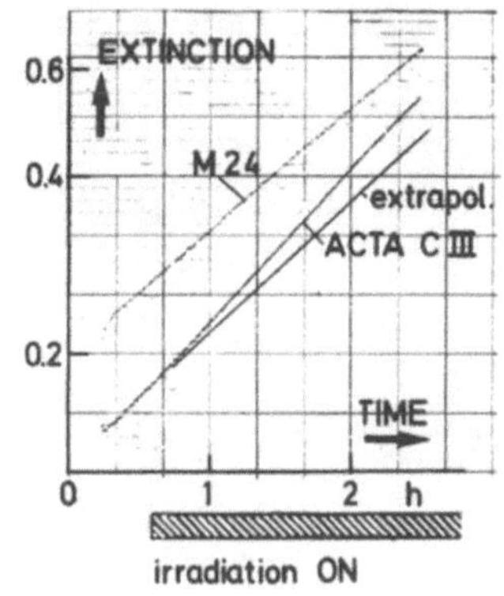

Fig.8.

Change of slope in the growth curve of a yeast culture induced by switching on irradiation at 41752 MHz (fork antenna, 21 mW absorbed power.

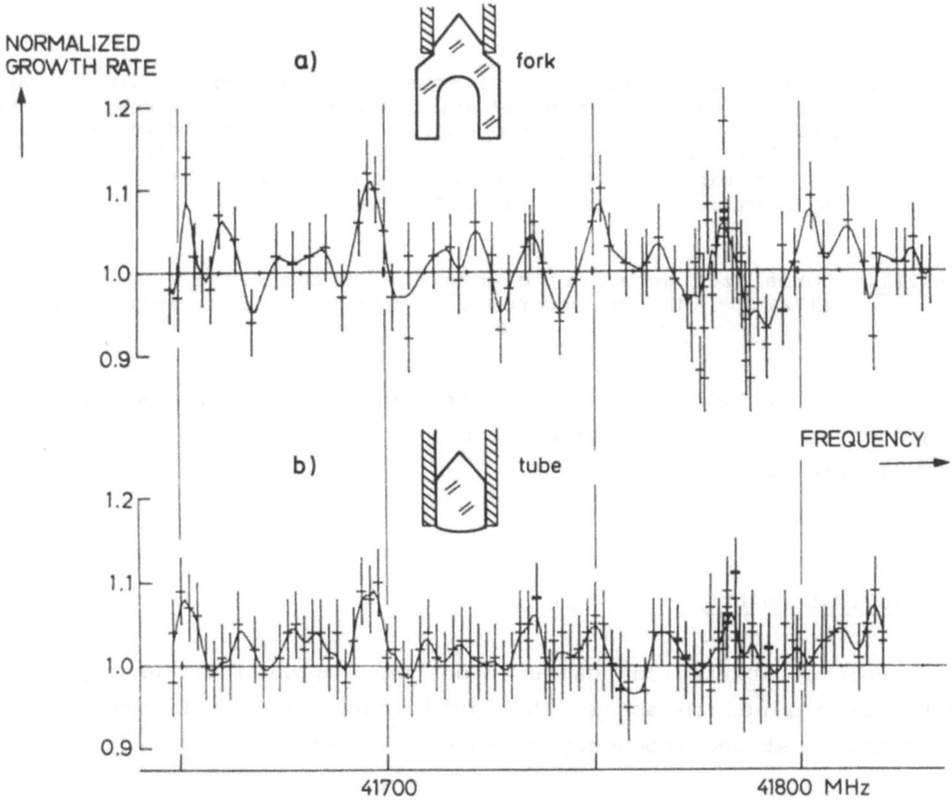

Fig.9. Normalized growth rate of yeast cultures vs. microwave irradiation frequency using (a) fork-shaped or (b) tubular antennae. The curves were obtained by single three-point smoothing.

These data prove that the microwave irradiation leads to fairly reproducible effects, increasing or decreasing the growth rate by up to about 10 %, depending on the frequency of irradiation. Further insight is revealed when the same growth rate data are replotted vs. the (absorbed) power which happened to be set for a given run:

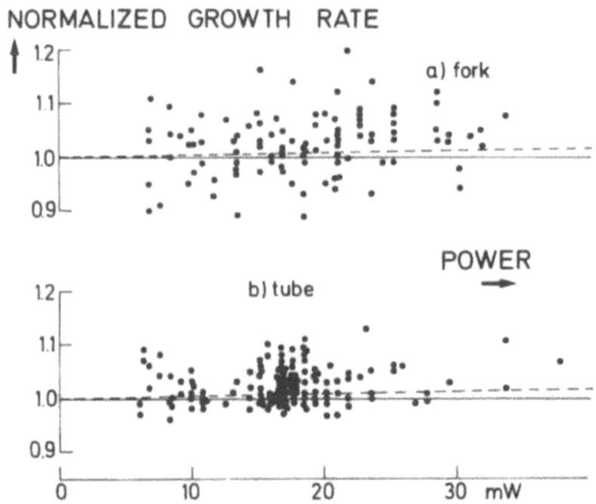

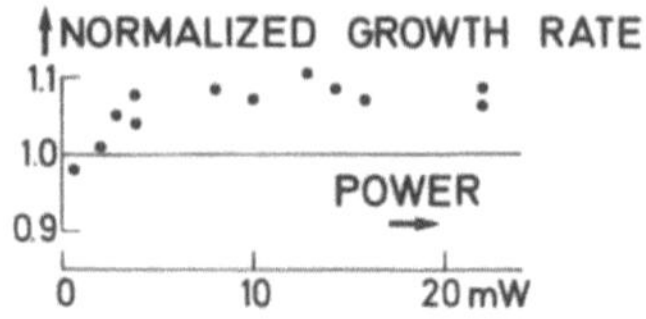

Fig.10. Normalized growth rate of yeast cultures vs. microwave power, irrespective of irradiation frequency (1:1 correspondence to Fig.9). For comparison dashed lines are drawn giving the magnitude of a thermal effect, as it is known from the temperature increase of the growth rate (0.027°C^{-1}) and the microwave-induced heating (0.016°C/mW).

Fig.11. Normalized growth rate obtained at the fixed frequency 41782 ± 1 MHz vs. microwave power using fork antennae.

Finally we replot seven fixed-frequency data points of Fig.9a, together with five additional ones obtained at power setting below 5 mW in Fig.11.

STATISTICAL ANALYSIS

Two types of statistical calculations were applied to our experimental data, both aimed to assess the significance of the important conclusion that a strong frequency dependence exists.

The first test calculates the probability that either spectrum in Fig.9a or Fig.9b could have been obtained as a mere chance sequence of a truly frequency-independent microwave effect. We first form a sequence of numbers $\bar{\mu}_i$ by arranging the ordinate values of e.g. Fig.9a in order of increasing frequency (a single average value is assigned where several data points share a common frequency). For this series we calculate the average (root mean square) difference of adjacent series elements, $d = \sqrt{\overline{(\bar{\mu}_{i+1} - \bar{\mu}_i)^2}}$. For comparison we then apply the same calculation to a large number of other series which are generated from the original one by a random perturbation of the sequence order. This yields a probability function of d which is graphically displayed.

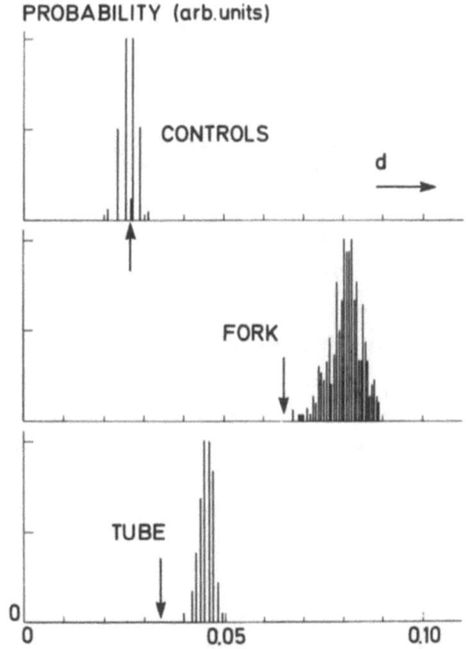

Fig.12. Probability distribution (histogram) and actual value (arrow) of average difference d between neighboring normalized growth rate values in experimental spectra. (a) control experiment series (Fig.4), (b) irradiation experiments using fork antennae (Fig.9a), (c) irradiation experiments using tube antennae (Fig.9b).

A relatively narrow probability distribution results for the average difference d of all three sets of experimental data in Fig.12. Only in case of the control experiments does the actual value d of the original series (arrow) fall within the central part of the probability distribution. For both irradiation experiments the actual value d lies well below the most probable value $\bar{d}$, by 3.6 and 5.4 standard deviations $s = \sqrt{\overline{d^2} - \bar{d}^2}$ for the fork and tube results, respectively. This allows the conclusion that the existence of a <u>frequency-dependent</u> microwave effect is ascertained for each of the experimental spectra Fig.9a and 9b to a significance level of $\alpha = 2.10^{-4}$ and 10^{-7}, respectively (from $\alpha = \frac{1}{2\pi} \int_z^\infty e^{-x^2/2} dx$ using $z = (\bar{d} - d)/s$).

The second statistical test is aimed to the question whether the frequency dependence of the observed microwave effect is the same for both experimental series Fig.9a and Fig.9b. A positive answer - as it is indicated by visual inspection - would exclude again that the observed frequency dependence could originate in standing wave resonances of the irradiating antenna. We calculate the cross correlation function

$$C(\Delta f) = (f_x - f_y + \Delta f)^{-1} \int_{f_y}^{f_x+\Delta f} [\bar{\mu}_a(f) - 1.01].[\bar{\mu}_b(f-\Delta f) - 1.02]\, df$$

where $\bar{\mu}_a(f)$ and $\bar{\mu}_b(f)$ are the experimental spectra of Fig.9a and 9b, respectively. Furthermore f_1= 41648 MHz, f_2= 41820 MHz, and (x,y) = (1,2) or (2,1) for $\Delta f > 0$ or $\Delta f < 0$, respectively. The numbers 1.01 and 1.02 substracted in the equation above are the mean values of $\bar{\mu}_a$ and $\bar{\mu}_b$, respectively. Error bars of the correlation are calculated from the ± 4 % error bars of the experimental data by standard error propagation. The result is shown in Fig.13.

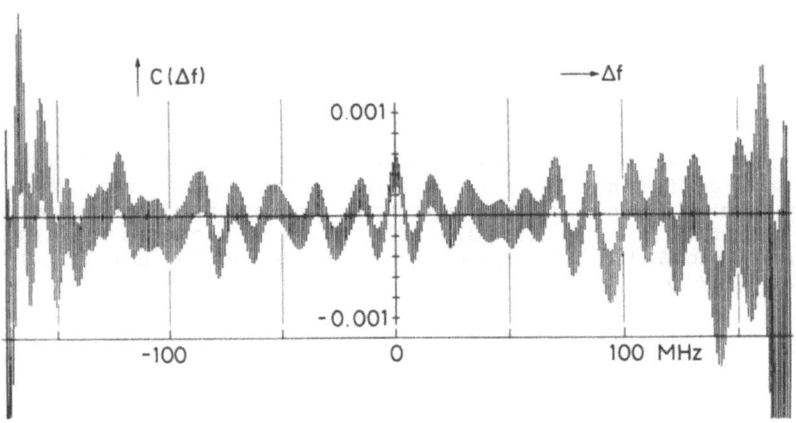

Fig.13. Cross correlation function between yeast growth rate spectra obtained with fork and tube antennae (Figs. 9a and 9b).

The result is a relative maximum of correlation with quite acceptable signal to noise ratio at zero frequency shift Δf indicating that indeed both spectra overlap.

It is furthermore of interest to apply a similar cross correlation test to compare the present results with those obtained in our previous study /5/. For this we use the interpolating curves of Fig.9b in /5/ and of Fig.9a in this work, disregarding error propagation.

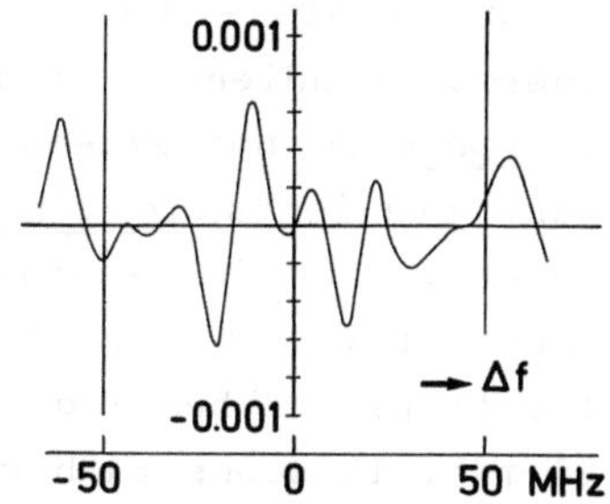

Fig.14. Cross correlation function of yeast growth rate spectra obtained with fork antennae previously /5/ and presently.

The result is that a relative maximum of correlation is obtained only at a nonzero frequency shift $\Delta f = -11$ MHz. Assuming that both spectra compared here in fact probe a common feature, i.e. a frequency selective response in the yeast cells, this result indicates that a systematic calibration correction of +11 MHz has to be applied to the frequency scale used in the previous study /5/. This correction would be reasonable since it is well within the ±20 MHz systematic error margin then estimated.

DISCUSSION

The existence of microwave effects on the yeast growth rate is convincingly proved by recorder traces as shown in Fig.8. These traces also show that half an hour of irradiation time after switching on the microwaves is sufficient for the growth rate change to take place. Further clear evidence for the microwave effect comes from the fact that the normalized growth rates with irradiation - disregarding for a moment the frequency dependence - very often lie outside the fluctuation limits of 1 ± 0.04 observed without irradiation (Figs.3,4) and range between 0.87 and 1.18, as is especially well demonstrated in Fig.10. In comparison, our earlier experiment /5/ had given somewhat larger negative effects down to $\bar{\mu} = 0.71$, for reasons which we do not understand yet but which could be due to the larger variation in temperature previously used.

The nonthermal nature of the observed microwave effect is directly demonstrated in Fig.10 where the dashed line gives the (measured) thermal effect produced by increasing the bath temperature. Caution -

in principle - has to be paid however to a possible influence of thermal gradients. In our experiment we have verified that gradients in space do not exceed 0.02°C/cm anywhere in the sample. On the other hand gradients in time of the temperature are given by the thermal inertia of the sample cuvette which is characterised by a time constant of 115 s, so that gradients do not exceed 0.005°C/s during the transient heating time after switching on the microwaves (of up to 40 mW). It seems rather unlikely that such small gradients should lead to biological effects.

When disregarding the frequency (Fig.10) the resulting rather arbitrary power dependence stands in sharp contrast to the threshold-like power dependence which was found by keeping the frequency fixed (Fig.11). This calls for a nonthermal origin of the effect. Similar power dependence was reported for millimeter microwave experiments with mice and *E. coli* /8/.

The fact that the power dependence for a fixed frequency comes out much clearer than the one which disregards the frequency already strongly indicates the existence of a frequency dependence. The statistical tests (Fig.12) on the individual spectra of Fig.9 clearly demonstrate the existence of a resonant microwave effect.

The correlation analysis of Fig.13 results in the statement that the

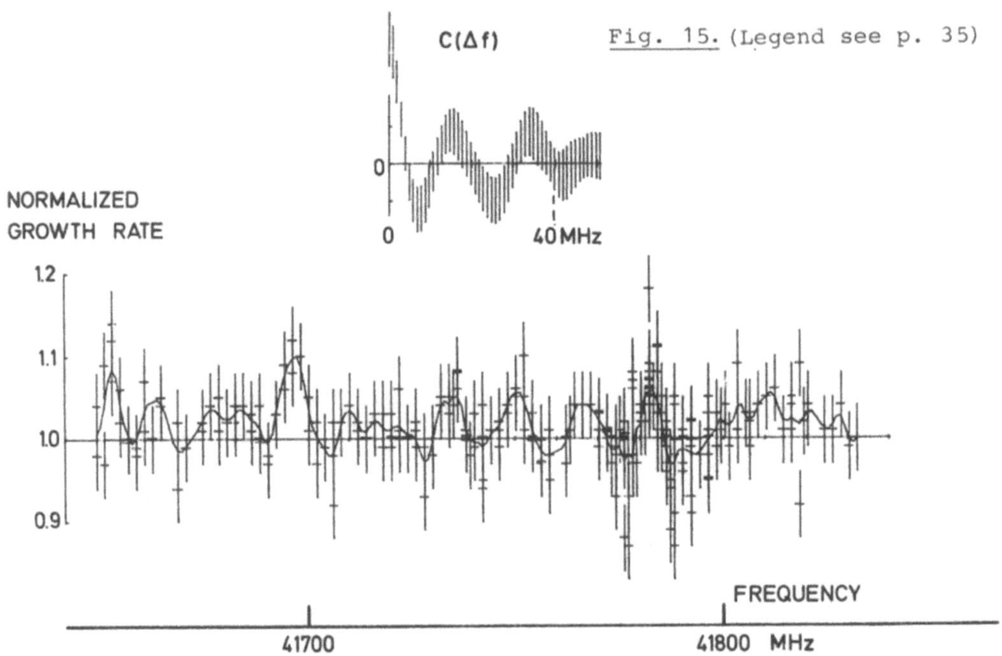

Fig. 15. (Legend see p. 35)

frequency dependence is reproducable using widely different irradiation geometries, which leads to the important conclusion that the resonance is inherent in the response of the yeast cells. This allows us to combine both spectra of Fig.9 in a single plot (Fig.15).

The autocorrelation function of the response spectrum (Fig.15) reveals interesting details of the average line shape of the observed resonances: On detuning from a prominent resonance the effect not only vanishes but readily builds up again with opposite sign until a detuning of 7 to 8 MHz is reached. Further detuning reverses the situation until a new resonance frequency is reached at a total offset of 16 MHz. Further satellites might also exist but become buried in noise. These observations are reminiscent of frequency modulation phenomena and thus may point to a possible role of a 16 MHz modulation frequency in the process.

A final remark concerns the observation that nearly equal size effects are found for both the tube and the fork antennae, although the latter has a 9 times larger teflon/water interface area so that a larger volume of suspension is irradiated. Let us try to estimate the effectively irradiated volumes for both antennae.

For this we assume that a threshold intensity I_c exists above which the microwave is effective. In case of a plane antenna surface of area A and total transmitted power P we obtain the effective irradiated volume $V_{eff} = (A/\alpha)\ \ln\left[P/(A.I_c)\right]$ where $\alpha = 45\ cm^{-1}$ is the power absorption coefficient of water. For a typical power P = 20 mW and assuming $I_c = 0.1\ mW/cm^2$ we obtain for the fork antenna $V_{eff} \approx 0.5\ cm^3$ but for the tube antenna $V_{eff} = 0.1\ cm^3$ only. The observation that the effects are nearly equal despite this difference can either mean that the critical intensity I_c is much smaller than the assumed 0.1 mW/cm², so that nearly all of the sample volume (2.5 cm³) is effectively irradiated. Another possible explanation has to do with the rapid mechanical stirring of the yeast suspension. We can assume that each cell reaches, at least once per minute any 0.1 cm³ sized volume element of the suspension. It is possible that the microwave effect is fully established for a given cell even when it is exposed briefly to intensities exceeding a certain critical one which are found only near the teflon surface.

Fig.15. Normalized growth rate response spectrum of yeast, obtained by combining all data in Fig.9. Insert: central part of autocorrelation function of this spectrum.

CONCLUSIONS

We have confirmed the existence of nonthermal resonant effects of 42 GHz irradiation on the growth rate of yeast suspensions. The resonances have narrow line widths (8 MHz FWHM) and are accompanied by sideband satellites offset by 16 MHz. In spite of the rather inhomogeneous intensity distribution within the suspension we have found strong indications for a threshold-like dependence on the irradiation intensity.

ACKNOWLEDGEMENTS

This study was supported by the Deutsche Forschungsgemeinschaft. We thank H. Fröhlich (Liverpool), L. Genzel (Stuttgart) and W. Pohlit (Frankfurt) for their support and collaboration. The technical assistance of I. Stark, W. Stark and O. Lock is gratefully acknowledged.

REFERENCES

1. H. Fröhlich, Int. Quant. Chem. 2, 641 (1968).
2. H. Fröhlich, Nature 228, 1093 (1970).
3. H. Fröhlich in Advances in Electronics and Electron Physics 53 (eds. L. Marton and C. Marton), pp. 85-152, Academic Press, 1980.
4. F. Kaiser in Biological Effects and Dosimetry of Nonionizing Radiation (eds. S. Michaelson, M. Grandolfo and A. Rindi) pp. 251-282, Plenum Press, New York, 1982. cf. also contribution in this volume, pp. 149-154.
5. W. Grundler and F. Keilmann, Z. Naturforsch. 33c, 15 (1978); W. Grundler, F. Keilmann and H. Fröhlich, Phys. lett. 62A, 463 (1977).
6. L. Genzel, F. Kremer, A. Poglitsch and G. Bechtold, contribution in this volume, pp. 69-85.
7. S.J. Webb and D.D. Dodds, Nature 218, 374 (1968); S.J. Webb and A.D. Booth, Nature 222, 1199 (1969).
8. N.D. Devyatkov, L.A. Sevastyanova, R.L. Vilenskaya, A.Z. Smolyanskaya, V.F. Kondrateva, E.N. Chistyakova, I.F. Shmakova, N.B. Ivanova, A.A. Treskunov, S.E. Manoilov, N.P Zalyubovskaya, R.I. Kiselev, V.I. Gaiduk, Yu.I. Khurgin and V.A. Kudryashova, Sov. Phys.-Usp.16, 568 (1974).
9. A.J. Berteaud, M. Dardalhon, N. Rebeyrotte and D. Averbeck, C.R. Acad. Sc. Paris 281, D843 (1975).
10. S.J. Webb, Phys. Lett. 73A, 145 (1979).
11. L.A. Sevastyanova, Vestn. Akad. Med. NAUK SSSR 2, pp. 65-88 (1979), (in Russian).
12. L.A. Sevastyanova, S.I. Potapov, V.G. Adamenko and R.I. Vilenskaya, Biological Sciences, No.6, 1969 (in Russian); S.I. Potapov, L.A. Sevastyanova and R.L. Vilenskaya, Biological Sciences, No.3, 1974 (in Russian).

13. E.B. Bazanova, A.K. Bryukhova, R.L. Vilenskaya, E.A. Gelvich, M.B. Golant, N.S. Laudau, V.M. Melnikova, N.P. Mikaelyan, G.M. Okhokhonina, L.A. Sevastyanova, A.Z. Smolyanskaya and N.A. Sycheva, Sov. Phys.-Usp.16, 569 (1974).
14. N.V. Abramova, Yu.V. Makeyev, and F.A. Tenn, Elektron. Obrab. Mater. (USSR) 2, 83 (1978), transl. in Electrochem. Ind. Process. Biol., pp. 74-75 (1979).
15. S.L. Arber, Elektron. Obrab. Mater. (USSR) 3, 59 (1978), transl. in Electrochem. Ind. Process. Biol., pp. 66-75.
16. N.P. Zalyubovskaya and R.I. Kiselev, Tsitologiya i Genetika 12, 232 (1978).
17. M. Dardalhon and D. Averbeck, Proc. IXth Int. Congr. Soc. Francaise de Radioprotection, Effets Biologiques des Rayonnements Non Ionisants, pp. 279-299, 1978.
18. M. Dardalhon, A.J. Berteaud and D. Averbeck, Radioprotection 14, 145 (1979).
19. F. Kremer, C. Koschnitzke, L. Santo, P. Quick and A. Poglitsch, contribution in this volume, pp. 13-24.
20. G. Nimtz, contribution in this volume, pp. 47-55.
21. C.F. Blackmann, S.G. Benane, C.M. Weil and J.S. Ali, Am. New York Acad. Sci. 247, pp. 352-366 (1975),
22. P. Tuengler, F. Keilmann and L. Genzel, Z. Naturforsch. 34c, 60 (1979).
23. D.L. Jaggard and J.L. Lords, Proc. IEEE 68, 114 (1980).
24. L.M. Partlow, L.G. Bush, L.J. Stensaas, D.W. Hill, A. Riazi, O.P. Gandhi, P.L. Inversen and M.J. Hagman, Biolelectromagnetics 2, pp. 123-159 (1981).
25. G. Larson and L. Karlander, Report TRITA-TET-8201, Royal Institute of Technology, S-10044 Stockholm (1982).
26. F. Keilmann, D. Böhme and L. Santo, Appl. Envir. Microbiol. 40, 458 (1980).
27. F. Keilmann, Collective Phenomena 3, 169 (1981).
28. W. Grundler, Collective Phenomena 3, 181 (1981).
29. W. Grundler, F. Keilmann, V. Putterlik and D. Strube, Br.J. Cancer 45, Suppl. V, 206 (1982).
30. F. Keilmann in Biological Effects and Dosimetry of Nonionizing Radiation (eds. S. Michaelson, M. Grandolfo and A. Rindi), pp. 283-298, Plenum Press, New York, 1982.
31. W. Grundler in Biological Effects and Dosimetry of Nonionizing Radiation (eds. S. Michaelson, M. Grandolfo and A. Rindi), pp. 299-318, Plenum Press, New York, 1982.
32. W. Grundler, F. Keilmann, D. Strube and I. Zimmermann, unpublished.
33. Nonthermal Effects of Millimeter Wave Irradiation, (ed. N.D. Devyatkov), Academ. Sci. USSR, Inst. Radiotech. Electrotech. Moscow, 1981 (in Russian).

On the Microwave Response of the Drosophila Melanogaster

GÜNTER NIMTZ
II. Physikalisches Institut, Universität zu Köln, D-5000 Köln 41

INTRODUCTION

Among the numerous investigations devoted to athermal biological effects induced by microwaves there are also a couple of experiments with the fruit fly (Drosophila melanogaster). The fruit fly represents a favoured and wellknown object of research in biology. This little fly is easily bred and the time of developing is rather short. Thus genetic effects can be studied over several generations within a couple of months.

At the beginning I shall report on data dealing with various athermal microwave effects in the fruit fly published recently in different countries. This will be followed by an extensive presentation and discussion of data obtained from experiments with the Drosophila melanogaster during the last two years in my group.

Why does a physicist start from his point of view such somewhat farfetched and obscure investigations? We are working intensively with microwaves. We apply microwaves in investigating transport properties of semiconductors, phase transitions of liquid crystals, and have developed microwave equipments for various commercial applications. Thus we were puzzled or even shocked by listening to the spectacular news about the hazardous athermal microwave effects in biological systems. Eventually we decided that we should carry out some investigations by our own in collaboration with biologists and physicians in order to look for these effects, which were unfortunately very often reported in a careless and unscientific way.

Coherent Excitations in Biological Systems
Ed. by H. Fröhlich and F. Kremer

ON DATA PUBLISHED UNTIL NOW

As far as I know the first investigation concerning the microwave response of the fruit fly was reported by Zalyubovskaya /74Z/. The experiments were carried out at frequencies between 37 and 55 GHz. The insects were irradiated either for 15 to 60 minutes or for 3 to 5 h. The radiation power was of the order of 10 mW/cm^2. Zalyubovskaya studied the influence of millimeter waves on the viability and on the ability to reproduce i.e. the fertility which is measured by the number of offspring. A lowered viability and fertility was observed to take place in the offspring of the irradiated adult insects. It was concluded that genetic changes were induced by the millimeter waves, since the lowered fertility was observed also in the second generation. The observed reduction in the fertility was up to 70% compared to that of the control insects. The biological effects of the millimeter waves were most pronounced at frequencies around 42 GHz. Unfortunately no figures of the statistical significance are given in this paper.

Dardal'hon et al. /77D/ investigated the response of the insect at 17 and 73 GHz. The radiation intensity was up to 100 mW/cm^2 at 73 GHz and up to 60 mW/cm^2 at 17 GHz. The insects were irradiated at different stations of their development, either for 3h as egg, or for 3h as larvae or for 2h as pupae. Also adult females were exposed to 17 GHz microwaves for 16 or 21h. The authors investigated viability, tumor incidence, and fertility, and concluded, that the results of viability and tumor incidence were not of high significance, however, the females irradiated at 17 GHz (60 mW/cm^2, 16h) and crossed with untreated males did exhibit a significant increase of fertility. They had typically 30% more offspring than the untreated insects.

Mutagenic action of microwaves were studied by Hamnerius et al. /79H/ at a frequency of 2.45 GHz. The power absorption rate was 100 W/kg which corresponds to an intensity of the radiation of about 200 mW/cm^2. Embryos 1 to 2h old in water were 6h irradiated. They used a sensitive somatic test system in which mutagenicity was measured as the frequency of somatic mutations for eye pigmentation. They observed 4 flies with mutations in 7512 irradiated insects and 2 flies with mutations in 3344 nontreated insects of the control group. According to this result it was concluded, that mutations were not induced by microwaves.

Similar conclusions using the same test on flies irradiated with 10 and 40 GHz at a level of 10 $\mu W/cm^2$ for 4h were drawn by Dennhöfer et al. /83D/ quite recently.

ON THE INVESTIGATIONS RUNNING AT COLOGNE UNIVERSITY

The following experiment was carried out in my group by Horst Aichmann in collaboration with the biologists L. Dennhöfer and C. Harte and the physicians A.-H. Frucht and H. Schaefer /83A/. The investigation was mainly devoted to the fertility of irradiated insects. Particularly the russian experiment on the fertility mentioned above should be repeated at the same microwave frequency.

We are studying the fertility of the Drosophila melanogaster (strain Berlin) for two years. Pupae are exposed to microwave intensities of about 10 $\mu W/cm^2$ at a frequency of 40 GHz for 120h. Calorimetric measurements have proved that a potential temperature rise can be ruled out, at least it cannot exceed 0.05^oC. The control insects are bred in the same environment except without microwave radiation. The experiments proceeded as follows: Two groups of pupae are taken from the same egg deposition. One group of them is irradiated for 120h in the pupae state as shown in Fig. 1. The other group is kept as control. A few hours after becoming an adult fly one female and two male insects from both groups respectively were crossed to start a family. From each group up to 15 such families are founded and their offspring is counted. The number of this offspring, i.e. the children of the irradiated generation and of the corresponding control insects represent the fertility of the P generation. In order to measure the fertility of the two successive generations F_1 and F_2 the previous procedure is repeated by crossing always one female and two males from the offspring of the irradiated and untreated insects respectively. The results of fertility of four different experiments (I to IV) are given in Table 1 for three generations P, F_1, and F_2, the first of these represents the irradiated generation. The main results are presented in Figs. 2 - 5, in Fig. 2 the relative deviation ($\bar{x}_{irr.}/\bar{x}_{contr.}-1$) of the arithmetic means of the irradiated and the untreated insects for the four experiments. There is a pronounced variation of the values of the four experiments. On the other hand there is obviously a tendency that the fertility of the first generation (P) is enhanced by the microwaves whereas in the third generation (F_2) a reduced fertility is observed. Two other presentations of the experimental data are shown in Figs. 3 and 4. The difference in offspring $\Delta = \bar{x}_{irr.} - \bar{x}_{contr.}$ is plotted in Fig. 3. In Fig. 4 it is shown how many families of an irradiated group have a higher or lower fertility as the average control family.

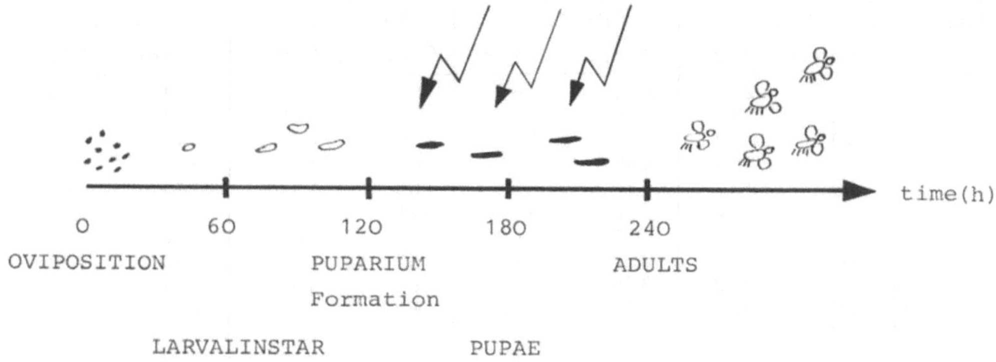

Fig.1. Dates of development for Drosophila melanogaster (T ≈ 25°C)

TABLE 1. The numbers of offspring of three successive generations P, F_1, and F_2 for irradiated and untreated insects. Data of four experiments are given which were finished I: October 1981, II: January 1982, III and IV: July 1982 respectively /83A/

Family	P(I)		P(II)		P(III)		P(IV)	
	$x_{contr.}$	$x_{irr.}$	$x_{contr.}$	$x_{irr.}$	$x_{contr.}$	$x_{irr.}$	$x_{contr.}$	$x_{irr.}$
1.	351	406	401	291	171	279	171	92
2.	21	496	219	637	170	290	285	286
3.	100	415	554	454	330	414	14	400
4.	105	595	0	557	573	516	273	317
5.	45	518	0	328	370	267	0	95
6.	84	343	392	361	573	389	168	430
7.	417	243	434	199	413	310	99	0
8.	641	576	247	362	144	319	429	254
9.	0		306	589	346	161	212	218
10.	295		38	370	415	351	470	262
11.	76		0					
12.	303		588					
13.			167					
14.			507					
$\bar{x}$	203	449	275	415	350	329	212	235
$\bar{x}_{irr.}/x_{contr.}$	2.21		1.51		0.94		1.11	
Δ(insects)	246		140		-21		23	
δ(families)	8		8		- 2		4	

$\bar{\Sigma}$ = 44% $\bar{\Delta}$ = 97 $\bar{\delta}$ = 18

Family	F_1(I)		F_1(II)		F_1(III)		F_1(IV)	
	$x_{contr.}$	$x_{irr.}$	$x_{contr.}$	$x_{irr.}$	$x_{contr.}$	$x_{irr.}$	$x_{contr.}$	$x_{irr.}$
1.	511	274	273	49	100	72	134	129
2.	318	349	46	436	75	128	74	82
3.	179	3	315	500	30	443	46	406
4.	1	294	585	386	83	126	312	53
5.	59	133	534	523	165	194	147	68
6.	99	292	96	117	385	134	272	140
7.	2	8	57	65	30	88	198	0
8.	308	63	257	669	62	83	194	113
9.	447	292	365	609	65	134	61	0
10.	600	223	377	413	83	593	214	67
11.	304		0					
12.	384		486					
13.			362					
14.			374					
$\bar{X}$	268	193	366	377	108	200	165	106
$\bar{X}_{irr.}/\bar{X}_{contr.}$	0.72		1.03		1.85		0.64	
Δ (insects)	-75		11		92		-50	
δ (families)	0		4		4		- 8	

$\bar{\Sigma}$ = 6% $\bar{\Delta}$ = -8 $\bar{\delta}$ = 0

Family	F_2(I)		F_2(II)		F_2(III)		F_2(IV)	
	$x_{contr.}$	$x_{irr.}$	$x_{contr.}$	$x_{irr.}$	$x_{contr.}$	$x_{irr.}$	$x_{contr.}$	$x_{irr.}$
1.	333	0	512	504	43	104	0	291
2.	503	0	461	49	315	4	0	259
3.	292	208	209	578	3	222	0	359
4.	103	0	401	549	294	18	0	24
5.	41	456	554	14	28	60	0	265
6.	31	3	0	459	288	12	12	85
7.	489	0	378	351	120	329	358	25
8.	407	233	440	448	0	304	495	0
9.	519	236	503	435	422	158	85	0
10.	0	490	231	275	0	61	160	0
11.			387					
12.			595					
13.			409					
14.			500					
$\bar{X}$	272	163	399	366	151	127	111	131
$\bar{X}_{irr.}/\bar{X}_{contr.}$	0.60		0.92		0.84		1.18	
Δ (insects)	-109		-33		-24		20	
δ (families)	- 6		2		- 2		- 2	

$\bar{\Sigma}$ = - 15% $\bar{\Delta}$ = - 36,5 $\bar{\delta}$ = -8

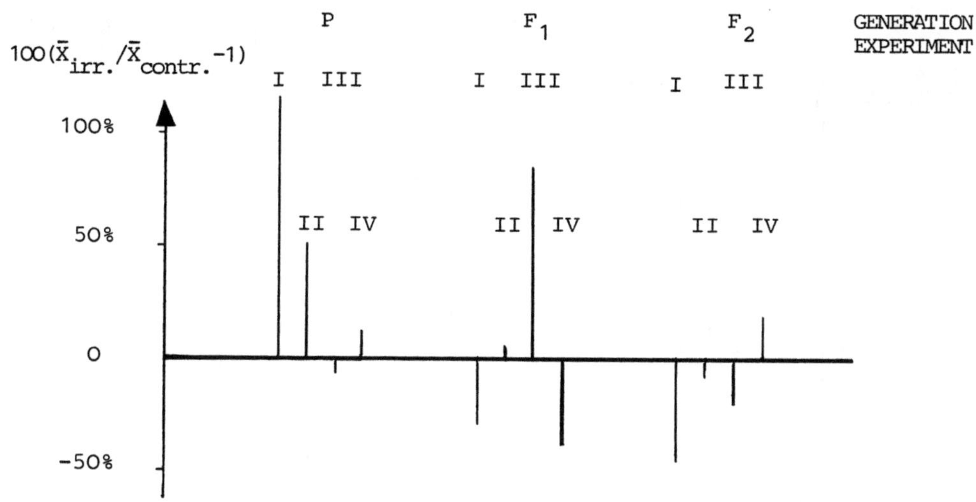

Fig. 2. Deviation of the fertility of irradiated insects P, of their children F_1, and of their grandchildren F_2 relative to the untreated insects.

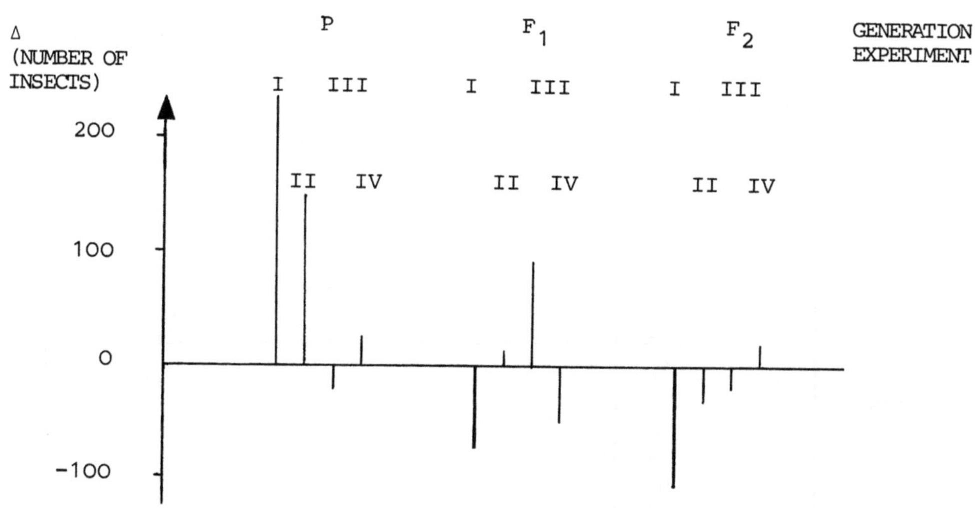

Fig. 3. Difference of offspring $\Delta = \bar{X}_{irr.} - \bar{X}_{contr.}$ between the arithmetic means of irradiated and untreated insects.

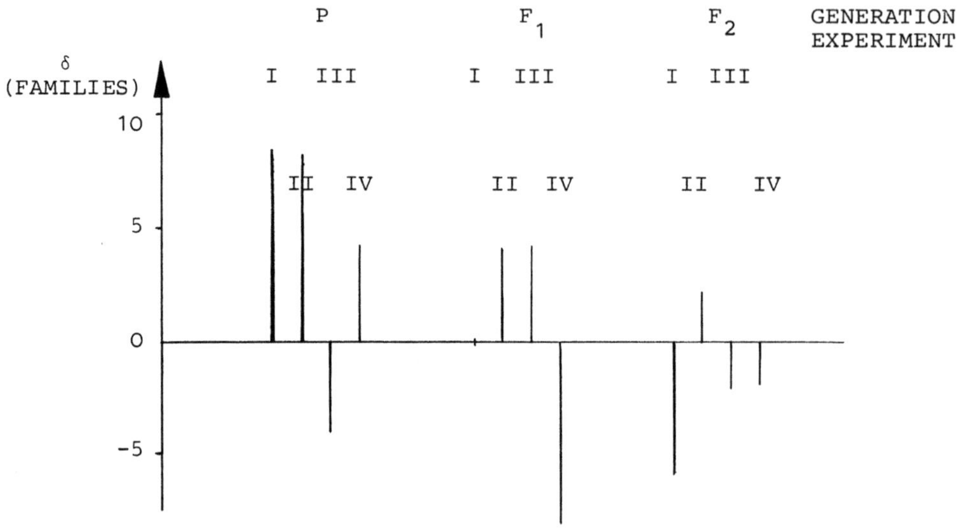

Fig. 4. Difference between the families of irradiated insects having more and those having less offspring than the arithmetic mean $\bar{X}_{contr.}$ of the untreated insects.

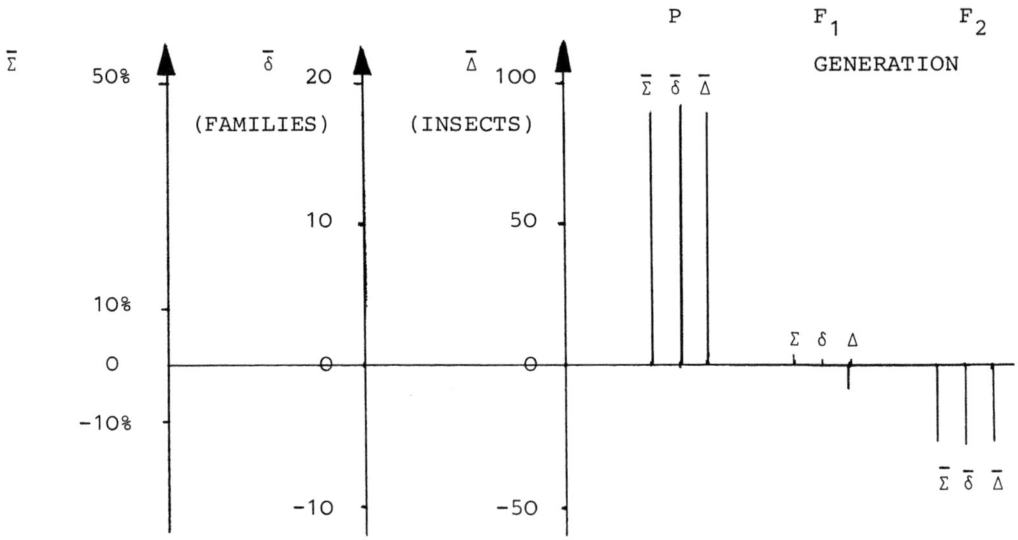

Fig. 5. The arithmetic mean of all four experiments of the properties shown in figs. 2 - 4.

We have defined

$$y_1 = +1 \quad , y_2 = 0 \text{ if } \quad (x_{irr.} - \bar{x}_{contr.}) > 0 \text{ and}$$
$$y_1 = 0 \quad , y_2 = -1 \text{ if } \quad (x_{irr.} - \bar{x}_{contr.}) < 0. \text{ With}$$
$$Y_1 = \Sigma y_1 \quad , Y_2 = \Sigma y_2 \text{ and } \delta = \Sigma (Y_1 + Y_2).$$

Both presentations, Fig. 3 and Fig. 4 display the same tendency for the fertility as that of Fig. 2.

The arithmetic means of the four experiments, defined as

$$\bar{\Sigma} = \frac{1}{n} \Sigma \; (x_{irr.}/x_{contr.} - 1), \; \bar{\Delta} = \frac{1}{n} \Sigma \; (\bar{x}_{irr.} - \bar{x}_{contr.}), \; \bar{\delta} = \Sigma \; \delta; \; n = 4$$

are presented in Fig. 5. All the three presentations reveal a strong increase in fertility in the first generation of irradiated insects. It seems that the second generation is not affected by microwave radiation whereas the third generation experiences a decrease in fertility.

CONCLUSION

The papers published so far have been indicative that athermal microwave effects may happen to the fruit fly. Criticism on these papers has been:

1) the high or undefined power level, which does not allow to exclude definitely thermal effects and mostly
2) the lack of statistical significance of the data. Natural systems seem to have a variability in magnitude a physicist is not used to.

Therefore we have applied an extremely low power level of 10 $\mu W/cm^2$ and decided to continue the experiments until we have obtained data enough for a decisive statistical interpretation.

The data obtained in four experiments are presented in Table 1 and Figs. 2 - 5. We have seen that the fertility of the P generation is strongly enhanced in the irradiated insects; their children, the F_1 generation, seem not to be affected, whereas the fertility of the grandchildren, the F_2 generation, shows a drop of some ten % compared to the fertility of the control insects. To get an idea of the statistical significance after four experiments we have carried out two statistical tests:

1. We carried out a Bernoulli trial and calculated the probability of finding Y_1 successes in $Y_1 + |Y_2|$ families according to a binomial

distribution with p = 1/2 for the experimental data presented in Fig. 4. For the P and the F_2 generations probabilities of $1.6 \cdot 10^{-4}$ and $5.7 \cdot 10^{-2}$ respectively were obtained.

2. In addition we applied the U-test of Mann-Whitney to the numbers of offspring given in the Table. Level of error for the hypothesis not to have equal fertility for irradiated and untreated insects was found to be 0.05 and 0.15 for the P and F_2 generations respectively.

Our experiment revealed a weak microwave induced biological effect at a low radiation power level of 10 $\mu W/cm^2$ corresponding to 60 mV/cm electric field strength. Our data are in qualitative agreement with those of Dardal'hon /77D/ who observed an increase of fertility in the P generation and with those of Zalyubovskaya /74Z/, who found a decrease of fertility at the F_1 and F_2 generations. For the time being we are performing two further experiments under the same experimental conditions in order to step up the statistical weight of our results.

The investigation is sponsored by the Berufsgenossenschaft für Feinmechanik und Elektrotechnik/Köln.

REFERENCES

74Z N.P. Zalyubovskaya, Sov. Phys. - Usp. 16, 574 (1974)

77D M. Dardal'hon, A.J. Berteaud, and D. Averbeck, Int. Symp. on Biological Effects (URSI), Airlie (USA), C2 (25. Oct. 1977)

79H Y. Hamnerius, H. Olofsson, A. Rasmuson, and B. Rasmuson Mutation Research 68, 217 (1979)

82D E. Dennhöfer, private communication (1982)

83A H. Aichmann, G. Nimtz, E. Dennhöfer, and C. Harte to be published (1983)

Effects of Low-level Millimeter Waves on Cellular and Subcellular Systems

S. M. MOTZKIN, L. BENES, N. BLOCK, B. ISRAEL, N. MAY, J. KURIYEL, L. BIRENBAUM, S. ROSENTHAL, and Q. HAN

Polytechnic Institute of New York, Brooklyn, NY 11201, USA

INTRODUCTION

Interactions of millimeter waves with living systems are believed to occur primarily on a subcellular or cellular level. Although most of the evidence that biological systems can be altered by low intensity electromagnetic fields is available primarily below 30 GHz, studies from the USSR and Eastern Europe (1,2) have indicated that above 30 GHz there are frequency dependent biological effects suggestive of resonance phenomena.

Our studies have been concerned primarily with the effects of low level exposures on colicin induction in E. Coli, and with mitochondrial calcium transport, and ATP production in oxidative phosphorylation. Although this paper will address in detail primarily the E. Coli investigations, I will summarize, briefly, the results of our mitochondrial studies.

MITOCHONDRIAL STUDIES

The effects of CW millimeter wave irradiation on oxidative phosphorylation and Ca^{++} transport in freshly prepared rat liver mitochondria have been examined at 8.58 mm (34.92 GHz) and at 41 discrete wavelengths between 5.0 and 6.0 mm (60-50 GHz). In oxidative phosphorylation ATP is synthesized by coupling the energy available from substrate oxidation to the thermodynamically unfavorable phosphorylation of ADP. The energy required in this reaction results from an electrochemical gradient created by a proton efflux during substrate oxidation. In the absence of ADP, substrate oxidation produces an energized state which in the presence of ADP is used in phosphorylation. Isolated mitochondria were evaluated for their efficiency in coupling succinate oxidation to phosphorylation of ADP. Oxygen consumption was followed with a Clark-type polarographic electrode. Tracings were used to calculate the respiratory control ratios at 30°, 25° and 4°C. Changes observed at 250, 500 and 1000 mW/cm^2 at 30°C indicated membrane damage. When the temperature was lowered (25°, 4°C) respiratory control ratios were maintained even at higher energy

Coherent Excitations in Biological Systems
Ed. by H. Fröhlich and F. Kremer

levels. As this technique does not enable detection of reaction reversal, an assay was developed which allowed simultaneous irradiation and analysis of ATP synthesis and could detect changes in energy coupling efficiency resulting from reversible membrane alterations. ATP was trapped as glucose-6-PO_4 and determined spectrophotometrically by observing changes in absorbance at 340 nm due, to ADP reduction. Changes in absorption are a function of the concentration of ATP in the sample. Exposures at 34.92 GHz and at .01-250 mW/cm^2 for 2 minutes and exposures at 41 discrete frequencies from 50-60 GHz at 5 mW/cm^2, all showed no effect. Increasing the irradiation time to 5 minutes did not alter the results.

Mitochondrial suspensions irradiated at 25°C, 8.58 mm (34.92 GHz) and 5.0-6.0 mm (60-50 GHz) at power densities of 1-1000 and 5 mW/cm^2 respectively, demonstrated no changes in $^{45}Ca^{2+}$ uptake after 15 minutes, or in efflux after 5 minutes of irradiation below 100 mW/cm^2. Above that power density decreased uptake and increased efflux were correlated with membrane damage. Alterations in mitochondrial function observed at high power densities may be attributed to the thermal burden on the sample.

E. COLI STUDIES

Smolyanskaya and Vilenskaya (2) have reported that millimeter waves at frequencies of 45.6 - 46.1 GHz (6.50 - 6.59 mm) and at 5.8 and 7.1 mm can stimulate colicin production. This effect is strongly frequency dependent (Fig. 1). Synthesis of colicin is enhanced to as much as an average of 300% at wavelengths of 5.8, 6.5 and 7.1 mm over a range of power densities from 0.01 to 1.0 mW/cm^2, and is almost insensitive to power density over 2 orders of magnitude.

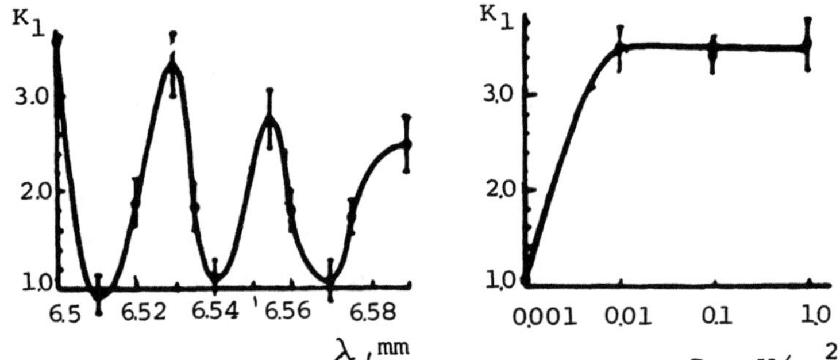

Figure 1.

RESULTS OF SMOLYANSKAYA AND VILENSKAYA

Attempts by many investigators to reproduce these experiments, or to carry out similar experiments relating to growth, have either been fruitless or have at best met with limited success.

Our primary goals have been to substantiate, if possible, reports of increased colicin induction in E. Coli, to clarify the question of frequency relationships and to achieve a better understanding of interactive mechanisms in the production of microwave effects. Millimeter wave exposure systems, in the 26.5 - 75 GHz band, using waveguide components which monitor power level, reflections and frequency, are being employed in single CW exposures at low power densities in experiments involving colicin induction in E. Coli.

INSTRUMENTATION

Experimental microwave exposure systems used are in the 33-50 (WR22) and 50-75 GHz (WR15) ranges. The 50-75 GHz system (Fig. 2) uses a Siemens RWO-60 backward wave oscillator powered by a Micro-Now 702 supply as the CW microwave source. At 50 GHz, the combination has a 2 hour frequency stability at least as good as 1 part in 5,000, after a 2 hour warm up period. Incident power is monitored on the branch arm of a directional coupler whereas reflections are measured with a slotted line. The power transmitted into the sample is determined from the incident power and a knowledge of the reflection coefficient. A rotary vane precision attenuator is used to adjust the power level in the waveguide; a wavemeter, with 0.02 GHz between adjacent divisions, is used to measure frequency, and a Hughes thermistor mount to measure power.

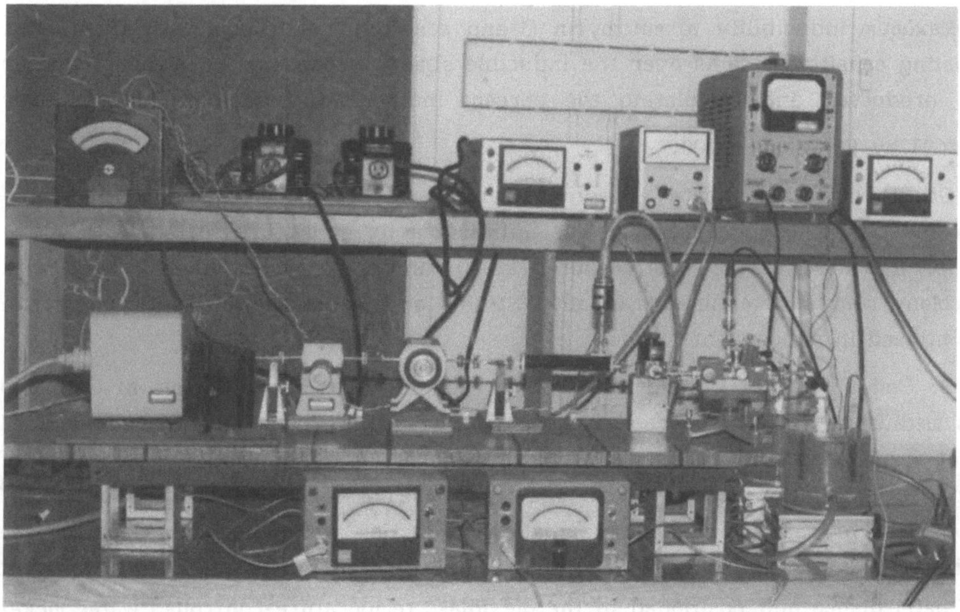

Figure 2.

The 33-50 GHz waveguide band also uses a Siemens RWO-60 backward wave tube. This system is powered by a Siemens RWON-14 power supply. In addition, the following components are utilized: a rotary vane attenuator to adjust the power level, Hewlett Packard Model 430B power meters, 2 Hughes temperature compensated thermistor mounts and directional couplers to monitor incident and reflected power.

Biological samples in a water-jacketed, temperature-controlled, plexiglass chamber are irradiated by using as the microwave applicator a Narda horn whose aperture is positioned about 2 mm above the sample surface. The specimen is prepared by depositing the E. Coli on a 0.2 mm thick millipore filter, which in turn is placed on a layer of agar in the petri dish. The combination is many wavelengths thick, so that power entering the surface is completely absorbed. The bacteria form a thin layer (~ 0.2 mm) on top of the filter. Since a skin depth is about 0.4 mm, their irradiation is relatively uniform. The waveguide instrumentation measures the net power leaving the horn. The assumption is made that all of it enters the medium directly below with a $\sin^2$ power distribution which mimics that in the horn aperture. Thus, the maximum power enters the medium at the center of the horn aperture and is equivalent to double the average power density. Reflections are measured by a standing wave indicator within the waveguide system. This data is then used to reset the variable attenuator so that the net power leaving the horn aperture remains fixed as the specimens are changed.

MATERIALS AND METHODS

Maximum inducibility to mitomycin C and sensitivity to colicin were determined by plating sensitive strains over the inducible strains, counting the number of clear areas produced, and calculating the percent increase of inducible (I) cells over shams(S) by $R = \frac{I-S}{S} \times 100$.

Growth characteristics of selected strains were determined for nutrient broth, Luria and Casamino acids media. The optical density of cultures grown with agitation for 8 hours at 37°C was measured at 30 minute intervals and correlated with cell colony counts determined manually. Strains and media showing optimum growth were utilized in subsequent studies.

E. Coli W3110 Col E_1 and W1485 were each inoculated into 15 mls of Casamino acids medium and grown overnight at 36-37°C (Fig. 3). The next morning a 1 ml aliquot was transferred to 15 mls of fresh medium. After approximately 3 hours of growth with agitation, at 37°C, to mid-log phase, a concentration of about 7×10^8 cells/ml was prepared for either irradiation, induction with mitomycin C or for use as shams. Five ml aliquots, filtered onto millipore filters, were placed into nutrient agar petri dishes and positioned in the plexiglass temperature-controlled water jacket directly underneath the aperture of the Narda horn. Shams and irradiated specimens differed only in their exposure to millimeter waves. Specimens were irradiated at

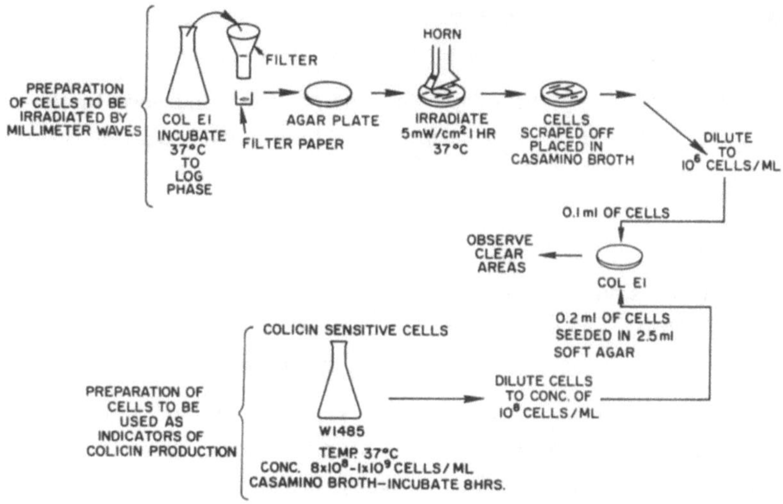

Figure 3. Protocol for Irradiation of E. Coli

discrete frequencies in 0.01 mm intervals between 5.77 and 5.83 mm (Series I, II, III) and between 6.50 and 6.59 mm for 1 hour at 37°C, and at power levels of 5, 0.5 and 0.05 mW/cm^2 (A,B,C respectively). A defined area of cells was removed from the center of the irradiated field and from an equivalent sham area and placed in fresh medium. Cell concentration was determined by spectrophotometric optical density measurement and the number of cells this represented determined from our standard growth curves. At this time the cell concentration 1-2x10^8 cells/ml, was diluted 100 times, and a 0.1 ml aliquot of these cells was spread evenly over a 90 mm diameter agar plate resulting in an approximate concentration of 1-2x10^5 cells/ml. Sensitive cells, W1485, transferred as described above for inducible cells, were grown for 4 hours to a concentration of 2x10^8 cells/ml. This was diluted 10x, and a 0.2 ml aliquot, a concentration of approximately 10^6 cells/ml, was seeded into 2.5 mls of soft agar which was then spread homogeneously over the Col E_1 cells. Plates were incubated at 37°C overnight. Clear areas produced were scored and measured to determine the number of cells induced and the quantity of colicin produced.

Random samples of bacteria, were treated with mitomycin C to determine inducibility. After 3 hours of growth, 1 µg/liter of mitomycin C was added to a flask of Col E_1 cells and grown with agitation for an additional hour. Treated cells were plated with sensitive cells as indicated above.

In addition, to determine the viability of the cells and to check the accuracy of the optical density measurements, an aliquot of irradiated specimens, randomly selected, was plated for colony counts.

Results obtained with irradiated or with mitomycin C induced specimens were plotted as a percent increase (or decrease) in colicin production over controls. Data was statistically evaluated to determine the significance of results obtained at a 95% confidence level. A Student t test was used to determine significance.

Each point on the graph represents the % increase in the number of clear areas of irradiated over control samples, as determined from the average of counts in three petri dishes, and corrected for the number of cells in solution.

RESULTS

Of the several strains tested for maximum inducibility and sensitivity W3110 Col E_1 and W1485 were selected for use in these experiments. The enhanced rate of growth in Casamino acids resulted in its selection as the medium of choice.

Utilizing these strains, three series of experiments have been carried out at 0.01 mm intervals in the 5.77- 5.83 mm range. In series IA at 5 mW/cm², a trend was evident which suggested a very broad peak at 5.78 mm of about 88% increase (Fig. 4). Statistically all data in this group is significant to the 95% confidence level. Preliminary experiments at the same wavelengths but at 0.5 mW/cm² (Series IB) suggest a trend which indicates a resonance type of pattern with peaks at 5.78, 5.80 and 5.83 mm (Fig. 5). Statistically, only 5.78, 5.81 and 5.83 mm appeared to be significant with the very small number of experiments available.

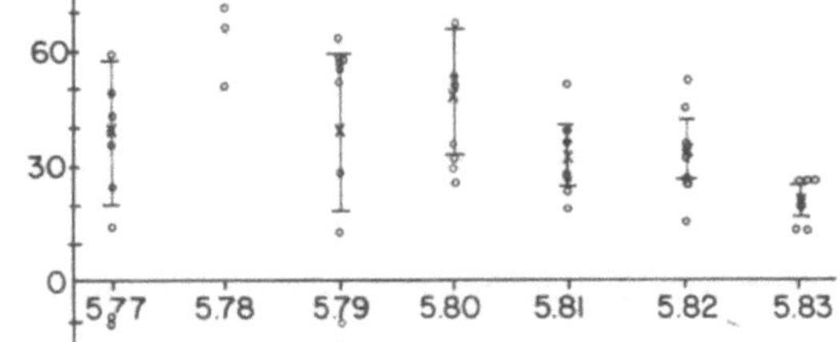

FIGURE 4.

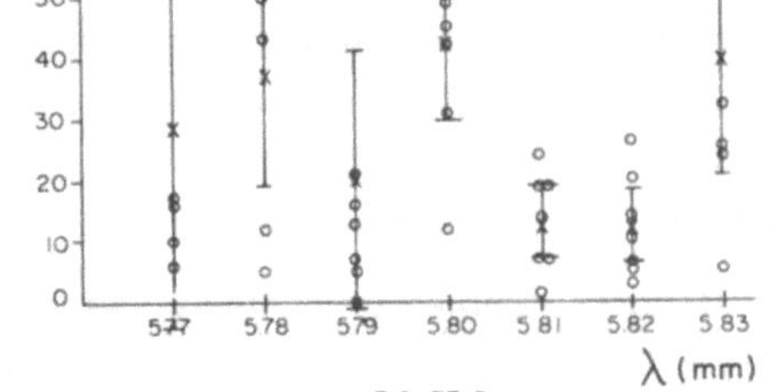

FIGURE 5.

Series II A, B, C, undertaken to determine the validity and reproducibility of experiments in Series I, used the same protocol at 5, 0.5 and 0.05 mW/cm². A smaller number of clear areas were observed consistently in Series II when compared with Series I and III. The standard deviation at any specific wavelength was very large (Figs. 6,7,8) and the results obtained were not statistically significant. Series III involved further attempts to clarify variations in the above results by replicating

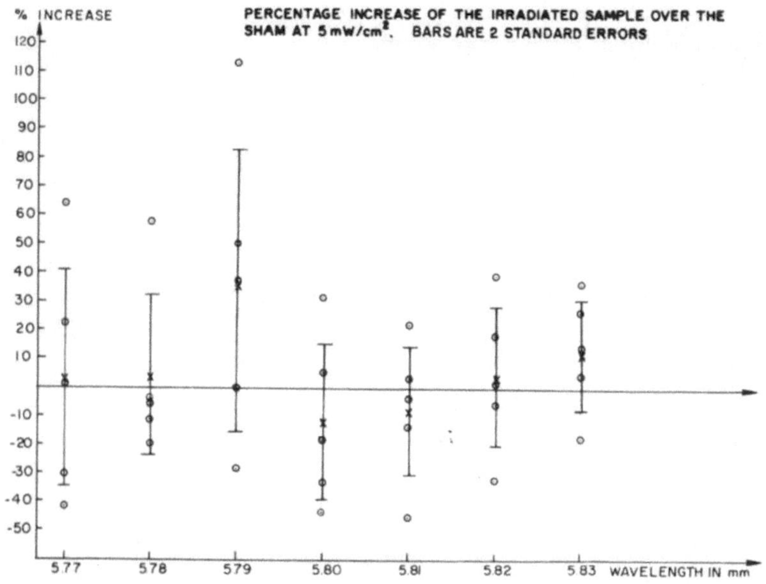

FIGURE 6.

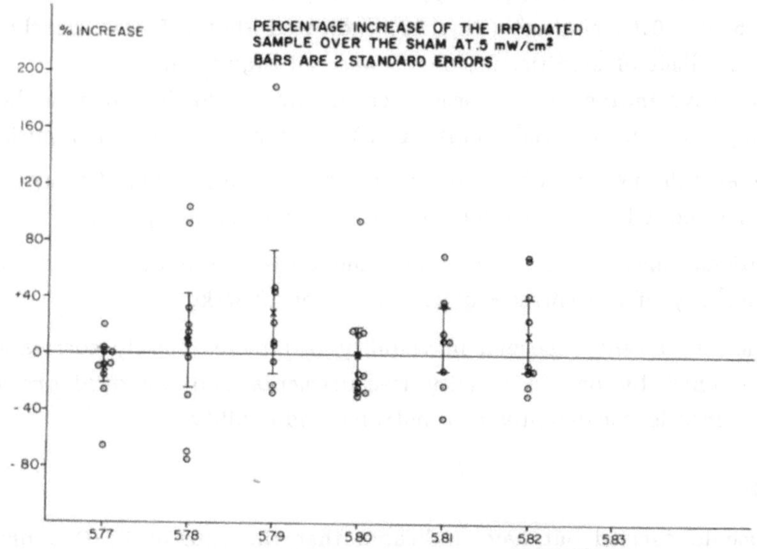

FIGURE 7.

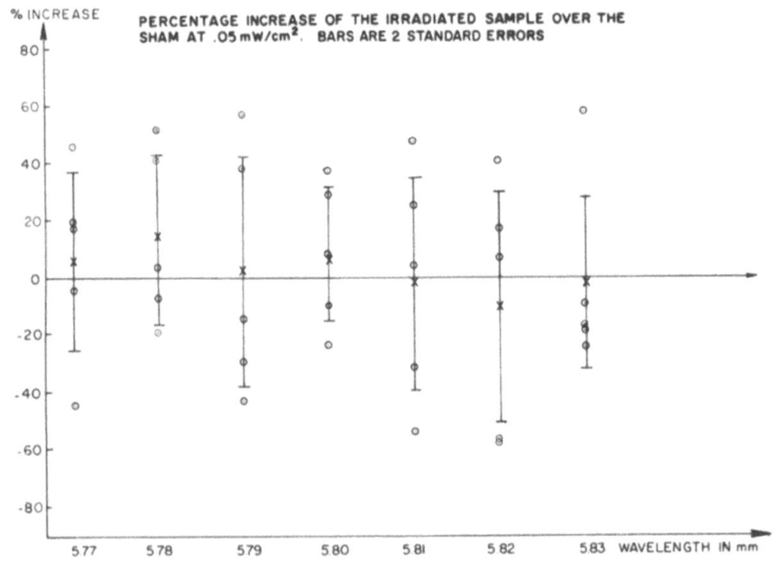

FIGURE 8.

the same protocol. In one replicate group significance was clearly noted at 5.77 mm and 5mW/cm^2 (Fig. 9). Subsequently, additional data taken at the wavelength 5.77 mm at 5, 0.5 and 0.05 mW/cm^2 (Fig. 10) indicated that 0.5 was clearly significant 0.05 was at the limit of significance and 5 was not significant.

Discrete wavelengths at 0.01 mm intervals were also examined in the 6.50-6.59 mm range at 5 and 0.5 mW/cm^2 (Figs 11,12). All results were insignificant except at 6.57 mm at 0.5 mW/cm^2 which is at the borderline of significance. Extensive variation in reproducibility is evident at all wavelengths and powers.

Calculations carried out to determine the SAR (averaged over one skin depth) at a power density of 0.5 mW/cm^2 gave a value of 12 W/kg.

The number of cells observed in viability test plates closely correlated with the number determined by optical density measurements and standard growth curves. Mitomycin C controls consistently demonstrated inducibility.

DISCUSSION

Experiments carried out have indicated that the rate of E. Coli growth is affected by the medium in which the bacteria are grown. Based on these studies Casamino acids was deemed to be the optimum growth medium. That optimum growth is associated with specific media has been known for some time. Casamino acids medium contains Casamino acids and thiamine hydrochloride (growth stimulating substances) perhaps accounting for the increased growth rate observed in this medium.

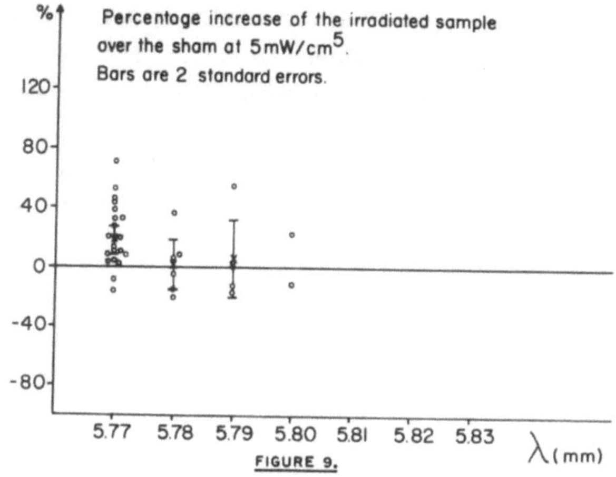

FIGURE 9.

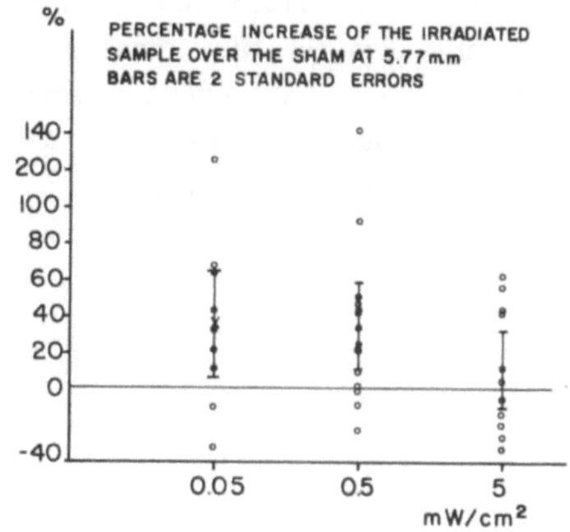

FIGURE 10.

Mitomycin C induction studies have indicated that, of the strains evaluated, W3110 Col E_1 and W1485 are the most inducible and sensitive strains respectively.

Frequency dependence of colicin inducibility, in strains selected, was examined between 5.77-5.83 mm and 6.50 -6.59 mm in cultures irradiated for 1 hour at 37°C at 5, 0.5, and 0.05 mW/cm^2. To date results observed in Series I, II, and III in the 5.77-5.83 and, 6.50-6.59 mm wavelength ranges indicate that irradiation with mm waves at discrete 0.01 mm intervals does not significantly stimulate colicinogenesis in the resonant fashion suggested by the Russians. However, there may be selected frequencies at which inducibility is achieved which is independent of power levels. Mitomycin C induction is statistically significant.

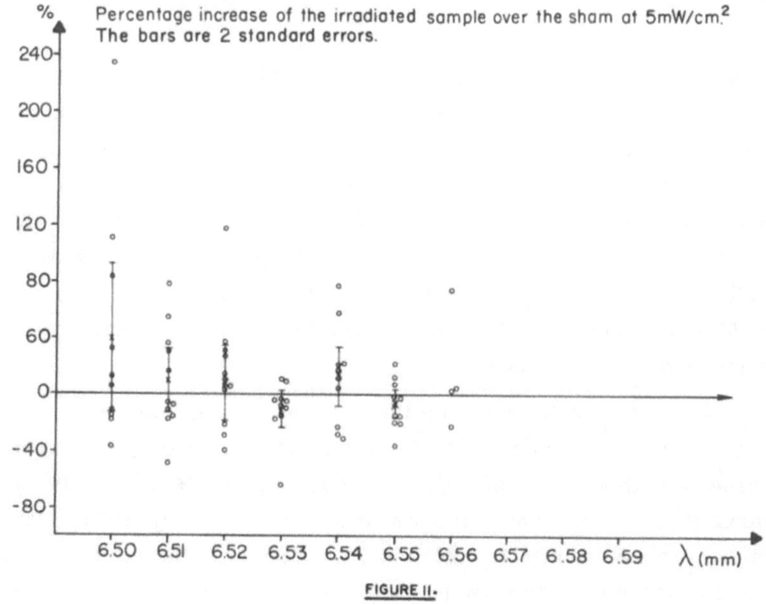

FIGURE II.

Results described above, in Series I, II, and III are inconsistent with each other and disagree with those reported by Smolyanskaya et al. (2). The latter have indicated that colicinogenesis is frequency dependent with sharp resonance peaks. In our experiments, Series IIA,B,C exhibit no peaks of any significance, whereas Series IA and IIIA suggest one broadband peak (Fig. 4) rather than a resonance pattern. Only in preliminary studies of Series IB (Fig. 5) was there a

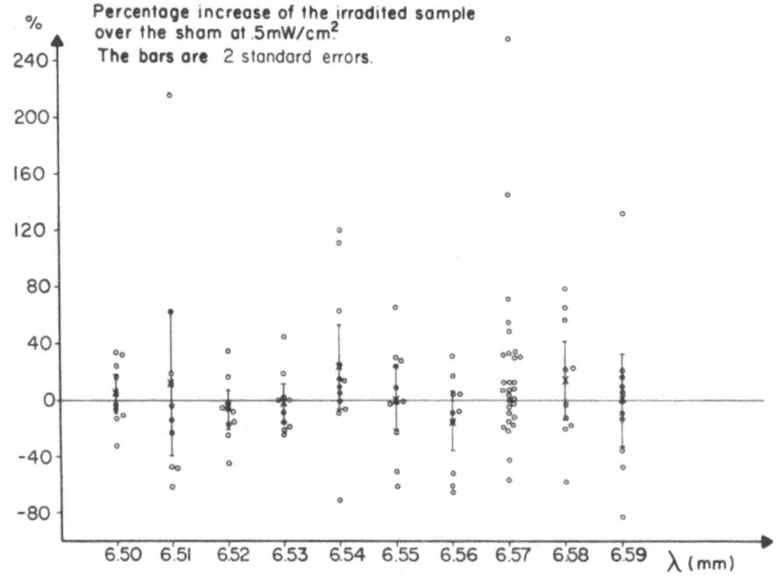

FIGURE 12.

suggested pattern of resonance similar to that reported by Smolyanskaya. Using this parameter there has been no consistently reproducible data substantiating the reports from the USSR. However, using growth parameters, a correlation with frequency dependence has been reported by other investigators, i.e., in E. Coli by Webb and Booth (3) and Berteaud et al. (4), in yeast, by Grundler et al. (5) and in Candida Albicans, by Dardanoni et al (6). In contrast with these results, in E. Coli, Hill et al. (7), Blackman et al.(8), Swicord et al. (9), Athey et al. (10) (also in lambda phage) and Kremer and Santo (personal communication) have not observed a frequency dependence.

Observed differences in the number of cells induced, and the random range of variability from induction to inhibition are not easily explained. It is unlikely that these are the result of different batches of bacteria. In some instances, limited numbers of preliminary experiments may yield significant data which subsequently appear to be incorrect. More likely however, the extensive variation observed may be due to the asynchronous nature of cultures used. Although cultures often are presumed to be synchronized in stationary phase, they are not, as cell death and cell division do occur. Transfer of these cells may initiate growth at different rates resulting in further asynchrony. In addition, as colicin production is purported to be dependent on both host chromosome and plasmid DNA, intimately associated with cell replication, then cells at different stages may react differently resulting in the variability observed. Induction may also best be achieved at other stages of the growth cycle. Perhaps as shown by Drissler in laser experiments resonance phenomena in E. Coli are prevalent in the stationary phase. Therefore, experiments are being undertaken with synchronous cultures at lag and stationary phase.

In addition, in related work we have modified a spectrofluorimeter which will enable us to monitor continuously, during the irradiation process, the fluorescence of molecules incorporated into the cell membrane. This system may help us to determine whether microthermal alterations induce conformational or other biochemical or biophysical changes, such as potential changes across the membrane. Furthermore, this technique will enable us to overcome or bypass limitations inherent in millimeter wave studies such as the depth of penetration due to attenuation. Also, as the fluorescence patterns are monitored during the irradiation process, it will eliminate the problem of transient reversible effects being missed.

BIBLIOGRAPHY

1. Scientific Session of the Division of General Physics and Astronomy USSR Academy of Science (Jan. 1973) USP Fiz Nauk 110, 452-469 (July 1973) also (1974) Sov. Phys. USP., v. 16, 568-579.
2. Smolyanskaya, A.Z. and R.L. Vilenskaya (1973) Scientific Session of the Div. of Gen. Phys. and Astronomy, USSR Acad. Sci. Sov. Phys. USP., v.16, 571-572.
3. Webb, S.J. and A.D. Booth (1969) Nature, 222, 1199-1200.
4. Berteaud, A.J., M. Dardalhon, and N. Reybeyrotte (1975) C.R. Acad. Sc. Paris 281:843-846.
5. Grundler, W. and F. Keilmann (1978) Z. Naturforsh, 33c: 15-22.
6. Dardanoni, L., M.V. Torregrossa, L. Zanforlin and M. Spalla, Abst. URSI Symp. on the Biological Effects of Electromagnetic Waves (Helsinki, Finland) August 1978.
7. Hill, D.W., A. Riazi, O.P. Gandhi, and M.J. Hagmann, Bioelectromagnetics Soc., 3rd Annual Conference August 9-12, 1981, Washington, D.C.
8. Blackman, C.T., S.G. Benane, C.M. Weil and J.S. Ali, (1975) Annals N.Y. Acad. Sci. 247, 352-365.
9. Swicord, M.L., T.W. Athey, F.L. Buchta, B.A. Krop, URSI Symposium on Biol. Effects of M. Waves, Finland, August 1978.
10. Athey, W. and B.A. Krop. 1980 Bioelectromagnetics Symposium Abstract #125.

Acknowledgement

This investigation was supported in part by the Office of Naval Research Contract N00014-77-C-0413 and in part by the Environmental Protection Agency contract 68-02-3457.

Millimeter-wave and Far-infrared Spectroscopy on Biological Macromolecules

L. GENZEL, F. KREMER, A. POGLITSCH, and G. BECHTOLD
Max-Planck-Institut für Festkörperforschung, D-7000 Stuttgart 80

ABSTRACT

Broadband measurements of the millimeter-wave and far-infrared absorption (10 GHz - 10^4 GHz) of anhydrons haemoglobin, lysozyme, keratin, poly-L-alanine and various polycrystalline amino-acids are reported. All measurements were extended over the temperature range from liquid helium to room temperature. For the millimeter range this was attained by using the novel oversized cavity technique. The frequency- and temperature-depedence of the millimeter wave absorption is described as due to three distinct relaxation processes on a picosecond time scale occuring in asymmetric double-well potentials. These processes are most probably assigned to the NH...OC hydrogen bridges of the peptide backbone. The far-infrared spectra are also discussed in connection with vibrations of such hydrogen bridges.

INTRODUCTION

Knowledge concerning the optical properties of biological macromolecules such as proteins in the submillimeter region is sparse (1). The situation is even worse at millimeter wavelengths where almost nothing is known about the dynamical processes which determine the dielectric properties of the biomolecules in this region. Any knowledge of this could add greatly to the understanding of the molecular dynamics, e.g. vibrations which involve larger parts of the macromolecule or relaxation processes. It would be of further interest if one could find absorption lines which correspond to the extremely sharp spectral features already observed in the growth rate of weakly microwave irradiated yeast cells (2) (see article of W. Grundler in this book). Parts of this article have been published elsewhere (3-5).

Coherent Excitations in Biological Systems
Ed. by H. Fröhlich and F. Kremer

EXPERIMENTAL

Various experimental reasons are responsible for the lack of information especially in the mm-region. There is first the low absorption of biomolecules in contrast to the high absorption of liquid water which prevents at present the possibility of obtaining useful results from aqueous solutions. There is second the difficulty of handling solid samples in the small-sized single mode waveguide- or resonator-systems which are conventionally used in the mm-region. Furthermore, temperature variation over a wide range causes problems due to the different thermal expansion coefficients of the samples and the wave-guide material. Finally quasi-optical systems using mirrors, lenses and horn antennae suffer from disturbing standing waves and frequency-dependent diffraction effects when coherent microwave sources are used. We have, therefore, employed the rather novel technique of the untuned, oversized high-Q cavity (6) for conducting the mm-wave spectroscopy. This method measures only the absorption of a sample inserted into a cryostat inside the cavity, but is essentially not disturbed by scattering and diffraction. The fact that the untuned cavity is a broadband device enabled us to carry out measurements in the frequency region from 40 to 150 GHz. A specially designed fused-silica cryostat allowed the temperature range from 4 K to 300 K to be covered. Further measurements at 10 GHz were made with the usual cavity perturbation technique. Additionally, far-infrared (FIR) absorption measurements have been made in the regime from 20 to 400 cm^{-1} by means of Fourier-transform spectroscopy.

In order to determine the absorption coefficient α of the material the untuned cavity method requires an assumed value for n, the real part of the complex index of refraction. For all samples under investigation, therefore, a value of $n = 1.6$ was used as it can be shown that a variation of n within limits of 0.2 will change the calculated α-value by less than 2 % only.

The study, presented here, deals with the dielectric and optical properties of anhydrous haemoglobin (HB), lysozyme (LY), keratine (KE), poly-L-alanine (PLA) and additionally with the polycrystalline materials L-alanine (Ala), glycine (Gly), valine (Val), leucine (Leu) and tyrocine (Tyr). The experiments with the untuned resonator require large amounts of sample material. The powdered chemicals were pressed into disc-shaped pellets of 50 mm in diameter and about 12 mm thickness which had then almost the bulk density. To prevent the absorption of water onto the sample material and to remove existing water, the materials were dried over P_2O_5 for one week before the preparation of the pressed samples, and during the measurements they were kept under either dry nitrogen or helium. In the cases of LY and KE additional drying procedures were used. LY was kept three days under 110°C and KE for the same time under 40°C. For FIR measurements, films of HB with thicknesses of about 40 μm were cast from aqueous solutions having a concentration of about 0.03 g/g. The polycrystalline layers of the various amino-acids used for the FIR spectroscopy had thicknesses between 50 and 80 μm.

RESULTS

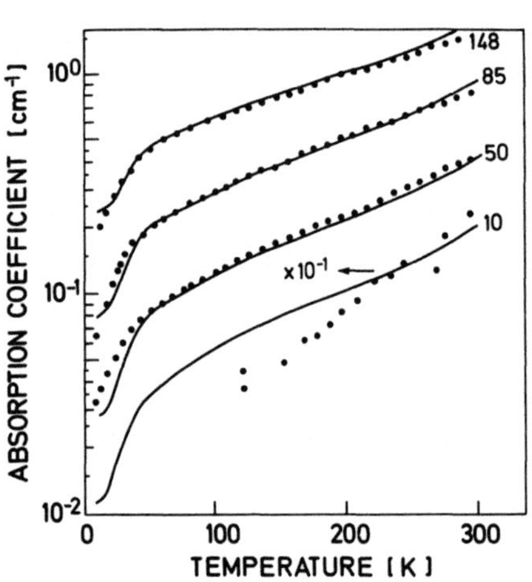

Fig.1.

Absorption coefficient vs. temperature of HB for the frequencies 10 GHz, 50 GHz, 85 GHz and 148 GHz. Dots are experimental results. The full lines are fitting curves according to Eq. 7 and table 1. Taken from (3)

The respective absorption coefficients α of HB, LY, KE and PLA are shown in Fig. 1 and Fig. 2. They all exhibit a similar temperature dependence, namely a nearly exponential increase of α with T from about 50 K to 300 K. Below 50 K, α dropped first rapidly with decreasing T and levelled off at about 10 K. Below 10 K the data are not accurate enough to draw further conclusions. The frequency dependence of α was found to be proportional to $\nu^{1.2} - \nu^2$. We did not observe any sharp features in the mm-wave spectra which could be discussed in terms of vibrational modes. For crystalline Ala at 70 GHz (Fig.3) values of α were found which are nearly one order of magnitude less than those of the biopolymers studied here. This measurement appears to be interesting because of the structure occuring in the T-dependence of α.

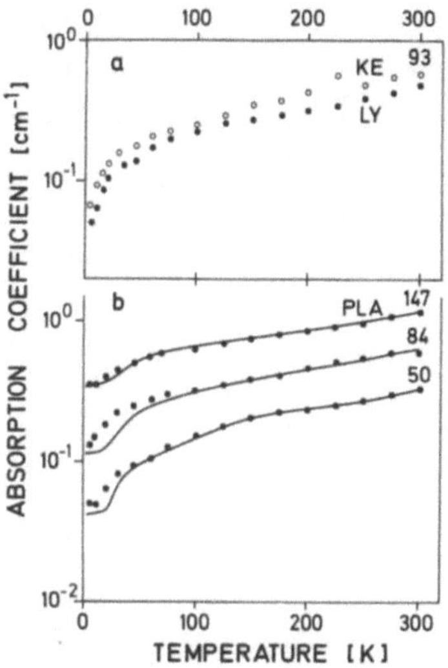

Fig.2.

a) Absorption coefficient vs. temperature (experimental) of KE and LY at 93 GHz.
b) Absorption coefficient vs. temperature of PLA for the frequencies 50 GHz, 84 GHz and 147 GHz. Dots are experimental results. The full lines are fitting curves according to Eq. 7 and table 2.

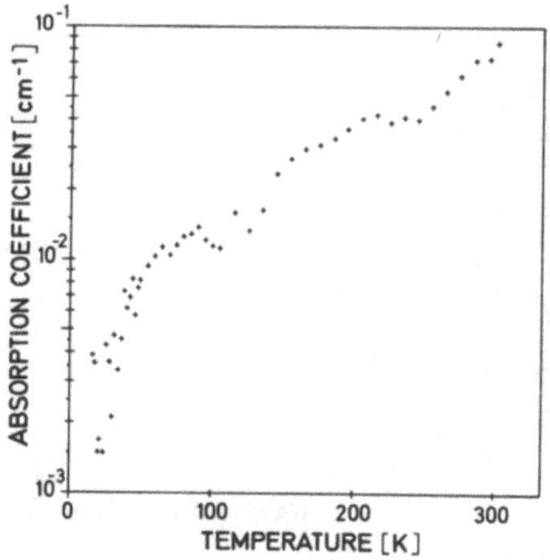

Fig.3.

Absorption coefficient vs. temperature (experimental) of polycrystalline Ala at 70 GHz. Taken from (3)

Several measurements of the FIR absorption of anhydrous proteins have been reported in the literature (5, 7-10). We present here only the spectral dependence of the absorption coefficient of HB between 20 and 500 cm^{-1} (Fig.4).

Transmission measurements were performed on a sample which was cooled to 6 K. It turned out that α is nearly temperature independent throughout the spectrum. All protein spectra studied so far are very similar with respect to the broad, unstructured peak around 140 cm^{-1}. Even the small band-like features which occur above 270 cm^{-1} are rather similar.

In order to have a better understanding of such protein spectra in the FIR we performed a series of investigations on polycrystalline layers of various amino-acids in the frequency range from 40-440 cm^{-1} and in the temperature range from 5 K to 300 K. The results for Gly, Ala, Val, Leu and Tyr are shown in Figs. 6-10 respectively (11). The spectrum of Ala has already been published by us together with its complete assignment (4).

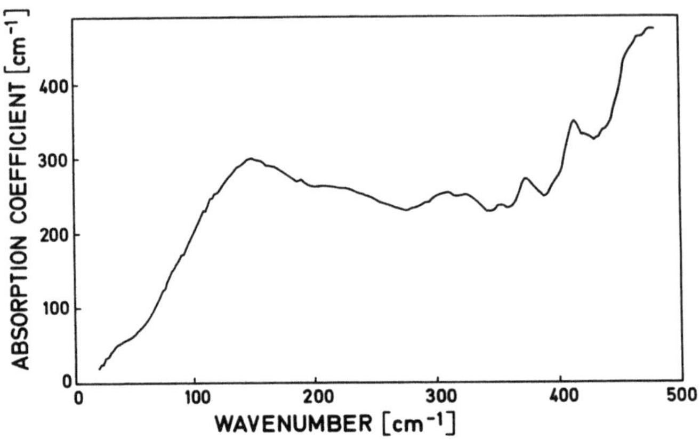

Fig.4. Absorption coefficient vs. wavenumber (experimental) of a HB-film at 6 K. Taken from (3)

DISCUSSION

Millimeter-wave spectra.

The most significant result of this study resides in the temperature- and frequency-dependence of the mm-wave absorption coefficient α of the various biomolecules and their respective similarity. We considered several known mechanisms which result in an increase of α with temperature such as multi-phonon difference processes, vibrational excitations from a thermally excited ground state, two-level resonant absorption as observed in amorphous materials, and dielectric relaxation effects. It turned out that only the latter processes are able to explain the observed frequency-dependence as well as the temperature-dependence of α, provided one uses an asymmetric double-well potential in which the relaxation occurs. To explain this we consider a Debye-type dielectric function

$$\varepsilon(\nu,T) = \varepsilon_\infty + (\varepsilon_0 - \varepsilon_\infty)(1 - i\nu\nu_\tau^{-1})^{-1} \quad (1)$$

where ε_∞ is the high-frequency dielectric constant (DC) and ε_0 the DC below the frequency regime of the relaxation process in question. ν_τ is the relaxation frequency which is usually assumed to have the following temperature-dependence for solids (12-14)

$$\nu_\tau = \nu_\infty \exp(-U/kT) \quad (2)$$

U is a potential barrier which has to be overcome thermally by the system in order to relax from one stable configuration to another one of the same energy implying, therefore, a symmetrical double-well potential model. It turns out that $(\varepsilon_0 - \varepsilon_\infty)$ of Eq. (1) is approximately given in this model by the Kirkwood-Fröhlich relation (12-14).

$$\varepsilon_0 - \varepsilon_\infty = 4\pi (kT)^{-1} \, \varepsilon_s (\varepsilon_\infty + 2\varepsilon_s)^{-1} \left((\varepsilon_\infty + 2)/3\right)^2 N\mu^2 \equiv CT^{-1} \qquad (3)$$

Here, ε_s is the static DC, N the spatial density of the double-well systems and μ is the absolute value of the vectorial difference of the dipole moments in the two wells. The absorption coefficient follows from Eq. (1):

$$\alpha = 2\pi\nu\varepsilon_2 (nc)^{-1} = (\varepsilon_0 - \varepsilon_\infty)(nc)^{-1} 2\pi\nu^2\nu_\tau (\nu^2 + \nu_\tau^2)^{-1} \qquad (4)$$

where n is the refractive index, c the light velocity and ε_2 the imaginary part of ε, Eq. (1). Eq. (4) with Eq. (3) can be discussed for two limiting cases: a) $\nu^2 << \nu_\tau^2$: In this case we find α to be proportional to $\nu^2\nu_\tau^{-1}\ T^{-1}$ yielding with Eq. (2) a strong decrease of the absorption coefficient with T and an increase with ν. b) $\nu^2 >> \nu_\tau^2$: Here, α is found to be proportional to $\nu_\tau T^{-1}$ and therefore an increase of α with T while the frequency dependence dropped out. It is evident that neither of these cases is in accordance with the experimental results (Figs.1 and 2) which makes it necessary to modify the model. Staying with relaxation processes, one is left with the use of an asymmetric double-well system as the next most simple model. In this, we assume that well 2 is energetically higher than well 1 by a potential V. Calling the depth of well 2 now U, the depth of well 1 will then be U+V. By treating this model one finds (3) instead of Eq. (3)

$$\varepsilon_0 - \varepsilon_\infty = CT^{-1} \exp(-V/kT)\left(1 + \exp(-V/kT)\right)^{-2} \qquad (5)$$

and instead of Eq. (2)

$$\nu_\tau = (\nu_\infty/2) \exp(-U/kT)\left(1 + \exp(-V/kT)\right) \qquad (6)$$

Inserting Eqs. (5) and (6) into Eq. (4), one can get the desired increase of α with T even in the case $\nu^2 < \nu_\tau^2$, retaining thus the frequency-dependence, if V > U. It turns out, however, that a single process of this kind is still not able to describe the observed $\alpha(\nu,T)$ over the whole range of variables in question. In such cases a continuous distribution of relaxation frequencies or of potentials V and U is often considered. In order to avoid the arbitrariness of fitting functions we used instead a minimal set of three relaxation terms which were able to fit the data. These three sets are also suggested from the result of the measurement on polycrystalline Ala (Fig.3) although one should expect for the case of biopolymers that really certain distributions around each of our parameter sets exist.

The observed levelling off of the $\alpha(T)$ curves at low T together with the observed ν^2-dependence can be explained by a temperature-independent wing of a strong and broad vibrational band found for proteins as well as for PLA in the FIR between 120 and 150 cm^{-1} (see Fig.4) which will be discussed in the following section. With this, our dielectric function for fitting the data of HB and PLA can be written

$$\varepsilon(\nu,T) = \varepsilon_\infty + \frac{S_o \nu_o^2}{\nu_o^2 - \nu^2 - i\nu\gamma_o} + \sum_{j=1}^{3} \frac{C_j}{T} \frac{e^{-V_j/kT}}{(1+e^{-V_j/kT})^2} \frac{1}{1 - i\nu/\nu_{\tau j}} \qquad (7)$$

As an example of the final fitting we show in Fig. 5 for HB the three relaxation processes and their sum to form the resultant $\varepsilon_2(T)$ curve. Table 1 and 2 compile the various fitting parameters used for the cases of HB and PLA, respectively, and the full lines in Fig.1 and Fig.2b show how well the experimental α (T,ν) data are represented.

The question arises: What is the microscopic nature of the assumed relaxation processes? Looking firstly on the values of the high-temperature relaxation frequencies ν_∞ (Table 1 and 2) one finds them all of the same order, resulting in rather short relaxation times of several picoseconds. This is indicative of relaxations in small molecular subunits. Secondly for the relaxation 3 the depth H = U+V of well 1 is 3,6 kcal/mol corresponding to about 1800 K. This value is not much smaller than typical binding energies of hydrogen bonds. We consider, therefore, relaxation 3 as being a disruption of H-bonds to form a weak Van der Waals bond.

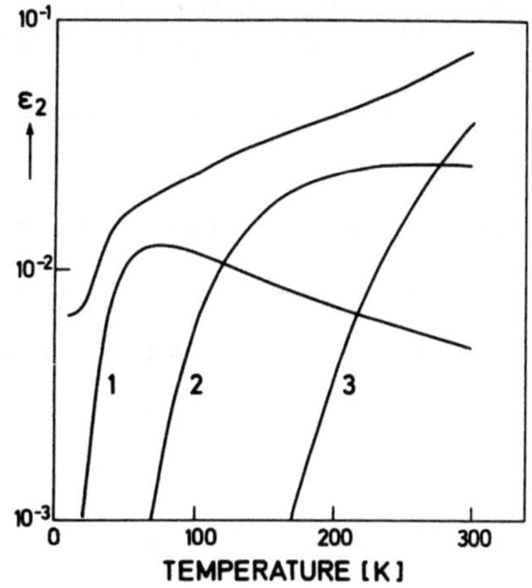

Fig.5.

The temperature dependence of the imaginary part ε_2 of HB at 85 GHz. Curves 1-3 correspond to the relaxation processes 1-3, respectively. The total dielectric function is composed of the 3 relaxation processes plus a vibrational term which causes the levelling off at low T (Eq. 7). Taken from (3)

Table 1. Parameters for fitting the mm-wave data of haemoglobin [Eq.(7)]. It is assumed that $N \approx 6 \cdot 10^{21}$ cm^{-3}.

Relaxation	ν_∞ (GHz)	U (kcal/Mol)	V (kcal/Mol)	$N\mu^2$ (cgs)	μ (Debye)
1	370	0.04	0.25	$2.8 \cdot 10^{-16}$	0.21 ± 0.01
2	350	0.20	0.91	$2.1 \cdot 10^{-15}$	0.59 ± 0.006
3	300	0.40	3.2	$8.1 \cdot 10^{-14}$	3.7 ± 0.4

FIR-vibration: $\varepsilon_\infty = 1.77$, $\nu_0 = 4400$ GHz ≈ 147 cm^{-1}

$S_0 = 0.385$, $\gamma_0 = 3879$ GHz

Table 2. Parameters for fitting the mm-wave data of poly-L-alanine [Eq.(7)]. It is assumed that $N \approx 9 \cdot 10^{21}$ cm^{-3}.

Relaxation	ν_∞ (GHz)	U (kcal/Mol)	V (kcal/Mol)	$N\mu^2$ (cgs)	μ (Debye)
1	230	0.04	0.25	$2.1 \cdot 10^{-16}$	0.15 ± 0.01
2	160	0.16	0.79	$1.5 \cdot 10^{-15}$	0.41 ± 0.06
3	260	0.40	3.2	$3.6 \cdot 10^{-14}$	2.1 ± 0.4

FIR-vibration: $\varepsilon_\infty = 1.77$, $\nu_0 = 3510$ GHz ≈ 117 cm^{-1}

$S_0 = 0.442$, $\gamma_0 = 3400$ GHz

There are, among others, two important and abundant groups of H-bonds in the polymers considered here. One group concerns the H-bonds of the NH...OC bridges which stabilize the peptide backbone. The other group belongs to the water molecules bonded by H-bridges to the amino-acid residues. We have performed experiments on hydrated samples of LY and polyamid and found that there occurs in both cases an extra relaxation at temperatures above 150 K which is different to the relaxations described so far. Therefore, we can conclude that the relaxation processes of our anhydrous samples are mainly due to the NH...OC bridges of the peptide backbone. From the values of the known molecular weight of the substances and their density and the known number of amino-acid residues per molecule, each of which yields just one NH...OC bridge, we get $N = 6 \cdot 10^{21}$ cm^{-3} for HB and $N = 9 \cdot 10^{21}$ cm^{-3} for PLA. This allows then the dipole moment change (μ in tables 1 and 2) to be calculated. For the relaxation 3 we arrive at $\mu = 3.7$ Debye for HB and $\mu = 2.1$ Debye for PLA. It is interesting to note that the NH...OC bridge itself has a dipole moment of 3.7 Debye (14). In this context the possibility can be considered that neighbouring bridges perform the relaxation in some cooperative way.

The discussion of the nature of the relaxation 1 and 2 must be more speculative. An indication that they might occur also on the NH...OC bridges is given by the fact that these bridges do not form a straight line. The oxygen atom with its local charge of about 0.38 e forms a dipole moment of the order of 1 Debye to the line connecting N and C.

One could, therefore, consider the relaxation 1 and 2 as being caused from asymmetric double-well potentials lying perpendicular to the H-bridge. It is worth noting that the observed relaxation effects, found here in solid and dry samples, are expected to exist also for the materials in aqueous solution as long as the submolecular units (H-bridges) in question are in the hydrophobic interior of the biopolymer and are thus not exposed to the outer water.

Far-infrared spectra

It was already pointed out above that the FIR spectra of various proteins are rather similar and that this is especially so for the broad and strong absorption band around 140 cm^{-1} (see also the article of J.B. Hasted in this book). It is the low frequency wing of this band which is considered to be responsible for the absorption in the millimeter range below 10 K (see Figs.1 and 2). The relatively low frequency of the band indicates weak force-constants of the underlying vibrational modes and/or large effective masses because of spatially extended eigevectors. It is known from other studies (15) that hydrogen-bond vibrations (stretch and bend) occur generally in the FIR between 40 and 180 cm^{-1}. One can, therefore, assume that the 140 cm^{-1} band of proteins is at least partly caused by vibrations of the peptide backbone with the weak H-bonds as force constants. This assumption is supported by an inspection of the amino-acid spectra (Figs. 6-10) being very rich in low-frequency vibrational bands which, in general split and sharpen up in going to low temperatures. It can be observed that the bands below about 200 cm^{-1} are sharper than the bands at higher wavenumbers. In the case of Ala we made a complete normal coordinate analysis including the lattice vibrations of the crystalline material (4). It is found that the lattice vibrations are strongly influenced by the NH...OC hydrogen bridges which bind the

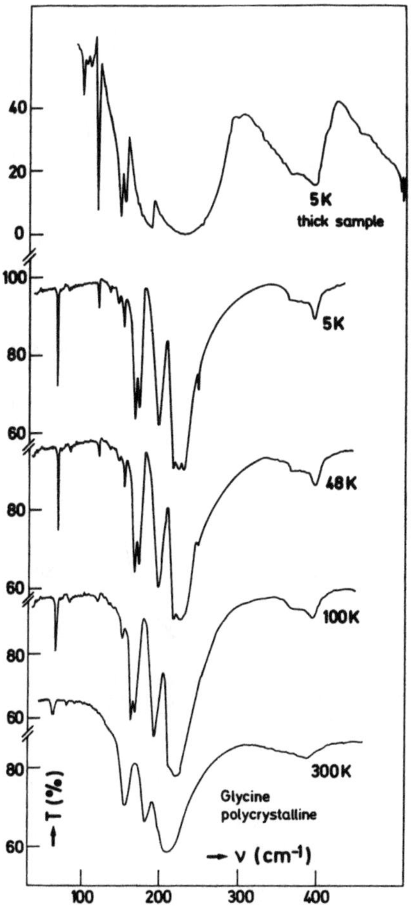

Fig.6. Transmission spectra vs. wavenumber of Gly at various temperatures

Ala-zwitterions in the lattice. The analysis showed that the sharp bands of Ala below 160 cm^{-1} are due to such lattice modes associated with stretch- and bending-vibrations of the hydrogen bridges. A similar situation is expected for the other amino-acid crystals although a complete analysis is not available at present.

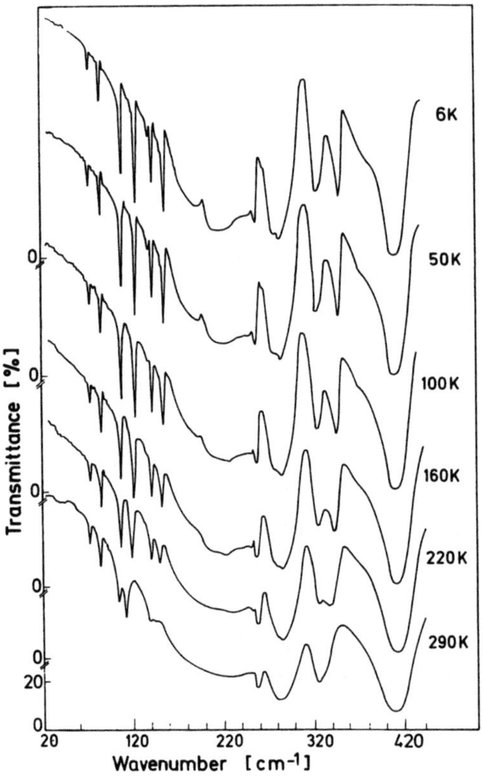

Fig.7.

Transmission spectra vs. wavenumber of Ala at various temperatures, taken from (4).

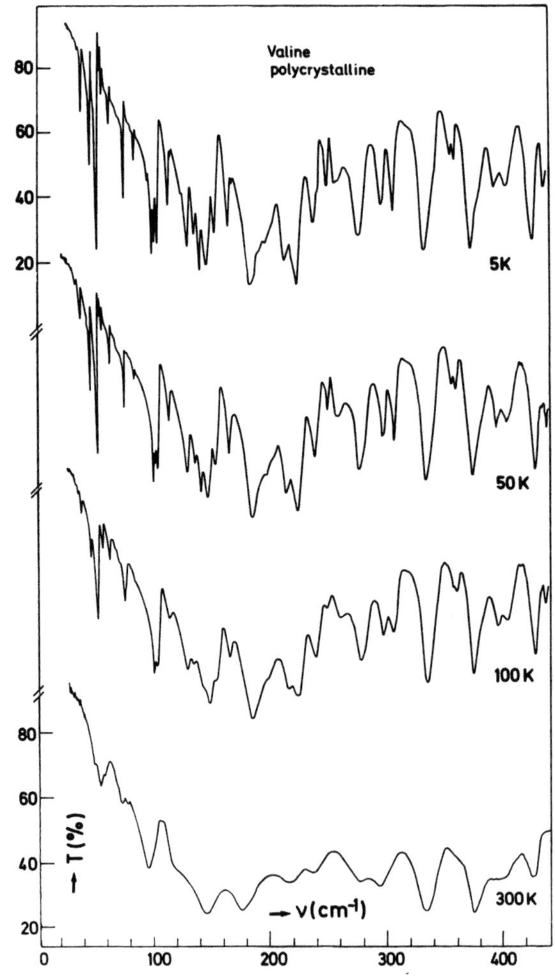

Fig.8. Transmission spectra vs. wavenumber of Val at various temperatures.

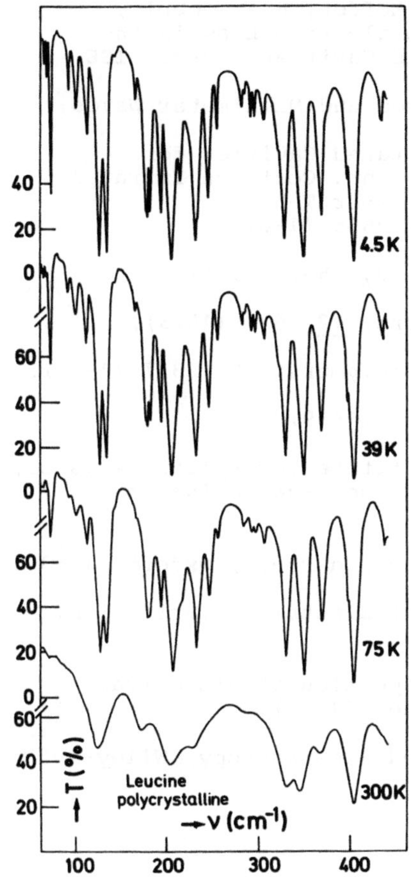

Fig.9.

Transmission spectra vs. wavenumber of Leu at various temperatures.

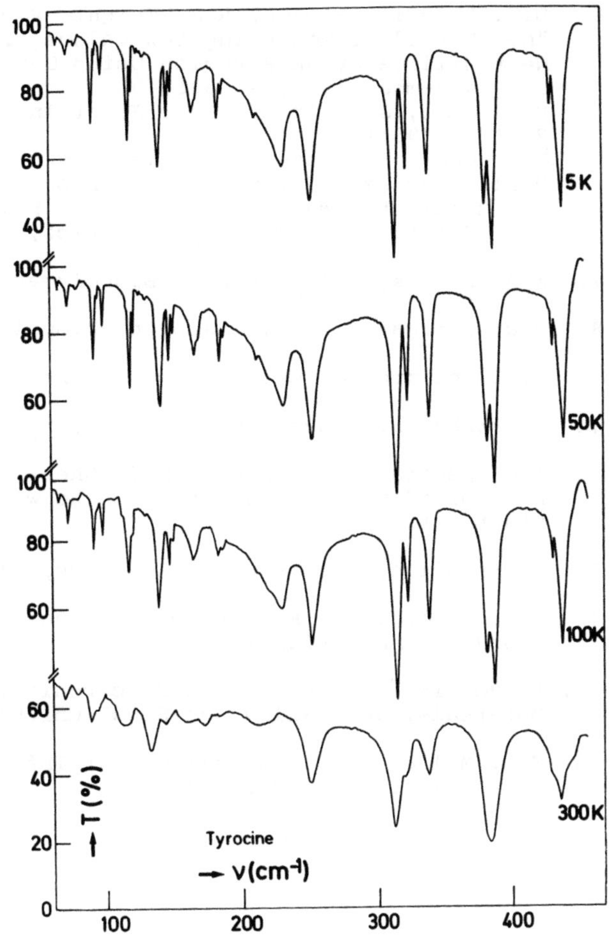

Fig.10. Transmission spectra vs. wavenumber of Tyr at various temperatures.

REFERENCES

1. G. Vergoten, G. Fleury, Y. Moschetto, in "Advances in Infrared and Raman Spectroscopy", Vol.4, 195-269. Ed. R.J.H. Clark, Heyden, London (1980).

2. W. Grundler, F. Keilmann, Z. Naturforschung 33c, 15 (1978).

3. L. Genzel, F. Kremer, A. Poglitsch, G. Bechtold, Biopolymers (1983) in print.

4. J. Bandekar, L. Genzel, F. Kremer, L. Santo, Spectrochim. Acta (1983) in print.

5. S.C. Shen, L. Santo, L. Genzel, Can. J. Spectrosc. 26 (3), 126 (1981).

6. D.T. Llewellyn-Jones, R.J. Knight, P.H. Moffet, H.A. Gebbie, "New Method of Measuring Low Values of Dielectric Loss in the Near Millimeter Wave Region Using Untuned Cavities", Proc. IEE, PtA, November 1980, P. 535.
F. Kremer, J.R. Izatt, Int. J. of Infrared and Millimeter Waves, 2, 675 (1981).
F. Kremer, L. Genzel, "Application of Untuned Cavities for Millimeter Wave Spectroscopy", Proc. 6th Int. Conf. on Infrared and Millimeter Waves, Miami, Dec. 1981, paper T-1-4.
J.R. Izatt, F. Kremer, Appl. Optics, 20, 2555 (1981).

7. W.J. Shotts, A.J. Sievers, Biopolymers, 13, 2593 (1974)

8. Y.N. Chirgadze, A.M. Ovsepyan, Biopolymers, 12, 637 (1973).

9. U. Buontempo, G. Careri, A. Ferraro, Biopolymers, 10, 2377 (1971).

10. M. Ataka, S. Tanaka, Biopolymers, 18, 507 (1979).

11. The spectra were taken by S.C. Shen, Institute for Applied Physics, Academia Sinica, Shanghai, when he was a guest-scientist in Stuttgart.

12. H. Fröhlich, Theory of Dielectrics, Clarendon Press, Oxford (1958).

13. C.J.F. Böttcher, P. Bordewijk, Theorie of Electric Polarization, Elsevier, Amsterdam (1973).

14. R. Pethig, Dielectric and Electronic Properties of Biological Materials, John Wiley and Sons, Chichester (1979).

15. K.D. Möller, W.R. Rothshild, Far Infrared Spectroscopy, Wiley-Interscience, New York (1971).

Excitation of Proteins by Electric Fields

J. B. HASTED[1], S. K. HUSAIN[1], A. Y. KO[1], D. ROSEN[2], ELISABETH NICOL[3], and J. R. BIRCH[3]

[1] Department of Physics, Birkbeck College (University of London), Malet Street, London WC1E 7HX, U.K.

[2] Department of Biophysics & Bioengineering, Chelsea College (University of London), 17a Onslow Gardens, London SW7 3AL, U.K.

[3] National Physical Laboratory, Teddington, Middx. TW11 OLW, U.K.

Introduction

We report measurements on the electrical absorption and dispersion of proteins, polysaccharides and their constituent aminoacids and saccharides, both at very low frequencies ($\nu = 10^{-3} - 10^{6}$ Hz) and in the submillimetre and far infra-red ($\tilde{\nu} = 10 - 550\ cm^{-1}$ (300 GHz - 16.5 THz)).

The submillimetre absorption spectra of the proteins and polysaccharides we have investigated show some characteristic structure, superposed on a background which is of similar shape to the liquid water spectrum (1). We attempt an interpretation as progressions from hydrogen bond vibrations, similar to those which we have found in crystalline aminoacids and mono- or disaccharides.

In the low frequency region, we have found (2) that the capacitance and loss angle of Langmuir-Blodgett multilayer preparations of haemoglobin show two field-dependent dispersion regions, near to 0.05 and 4×10^{5} Hz respectively. At sufficiently high fields these move closer together, forming a broadened relaxation at about 100 Hz; this state of the material persists for up to six weeks after the removal of the field, but can be destroyed by heat: it is essentially a metastable configuration. The electrical properties of denatured haemoglobin films can be permanently changed by the application of high fields.

Coherent Excitations in Biological Systems
Ed. by H. Fröhlich and F. Kremer

Submillimetre Wave Absorption Spectra

The submillimetre wave absorption spectra were determined by the techniques of transmission Fourier Transform Spectroscopy using a modular Michelson interferometer (3). The measurements were made in three ranges, 5-55 cm^{-1}, 30-150 cm^{-1} and 130-550 cm^{-1}. The overlaps enable us to estimate the levels of random and systematic error in the measured spectra. All spectra have been measured at laboratory temperature.

We have examined proteins both in aqueous solutions and in two solid forms - films cast from aqueous solutions and finely powdered crystals 20× diluted with fine non-absorbing polyethylene powder and compressed into a 16 mm diameter, 1 mm thick disc under weight. The compressed discs are superior as targets to cast films provided that the powder is sufficiently fine. They have smaller absorption coefficients and, essential for Fourier Transform Spectroscopy, their surfaces are plane parallel. They are maintained at 55% relative humidity when not in the spectrometer. Cast films can be more quickly hydrated in humid atmospheres, but during preparation their free surface forms a non-planar meniscus and they have a tendency to crack. The cast films have therefore only been used below 100 cm^{-1}, for variable hydration data.

Figure 1 shows the spectra of lysozyme, haemoglobin and serum albumin. Displayed with the lysozyme spectrum is the spectrum of liquid water (1); its characteristic broad 180 cm^{-1} band and shoulder of the broad 580 cm^{-1} band are comparable in shape to this and other protein spectra. The hypothesis that the bound water contribution dominates the protein spectra in this waveband is attractive; support is lent to it by measurements made on cast films of lysozyme between 5 and 100 cm^{-1} over a wide range of controlled humidities (4). The wing of the 180 cm^{-1} band is directly dependent upon the proportion h of bound water, determined by weight. The $\alpha(h)$ data at 55 cm^{-1}, displayed in Figure 3, bear out the bound water background hypothesis.

However, it is clear from Figures 1a,b and c that there are small structures superposed on this background. At low temperatures these structures are sharper, as has been shown by the Stuttgart group (5). The positions found by the Stuttgart group of the low temperature haemoglobin lines have been reproduced in Figure 1b for comparison.

We have measured a number of crystalline amino-acid spectra and it is apparent that there are sharp bands corresponding to parts of the weak protein

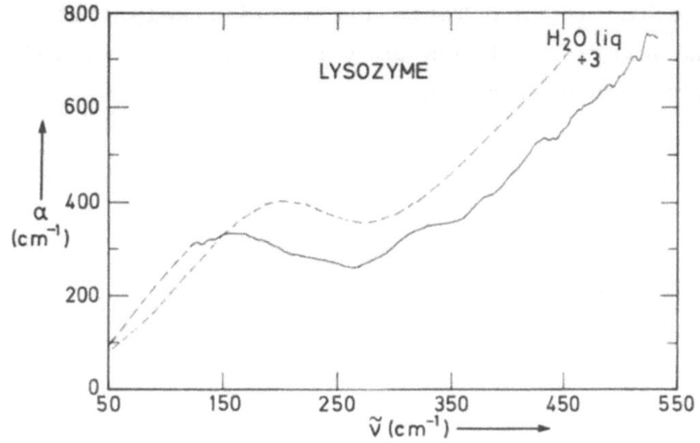

<u>Figure 1.</u> Absorption spectra of proteins and aminoacids.
<u>1a.</u> Lysozyme. Full line represents computed data, extended by broken line traced approximately from a lower frequency data run. The other broken line represents liquid water data (1) divided by a factor of 3

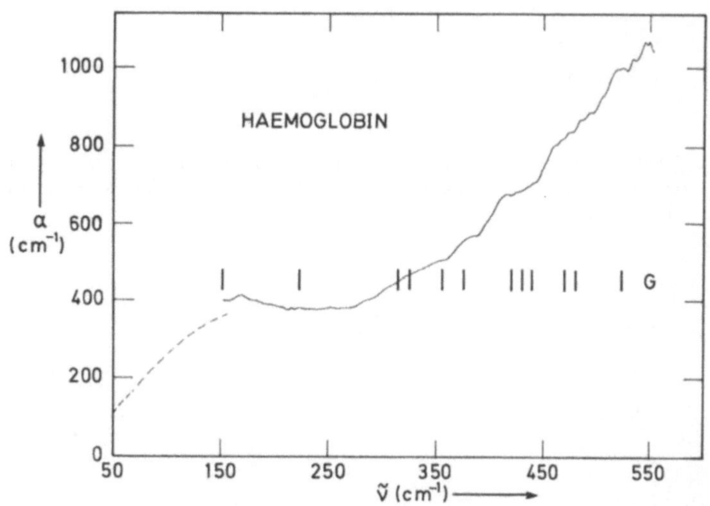

<u>1b.</u> Haemoglobin. Full line represents computed data, extended by broken line traced approximately from a lower frequency data run. Vertical lines represent bands appearing in the data of Shen et al. (5)

structure; in Figure 1c the tyrosine spectrum has been overlaid on that of serum albumin, and the bands at 320, 380 and 430 cm^{-1} may possibly be contributing to the structure in the protein spectra. However, it is not to be expected merely that selected aminoacid bands would show through in the protein structure; rather, progressions in the biopolymers and in the monomers are likely to be similar. We present these spectra in Figures 1c, 1d, and Table 1. It is likely that they contain NHO and OHO bond characteristic bands as well as lattice vibrations and contributions from intramolecular modes.

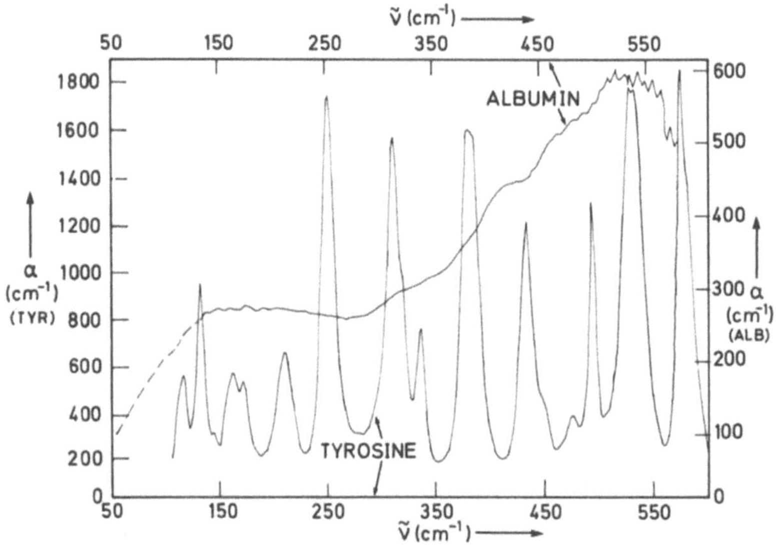

1c. Tyrosine and serum albumin. Full line represents computed data, extended by broken line traced approximately from lower frequency data runs

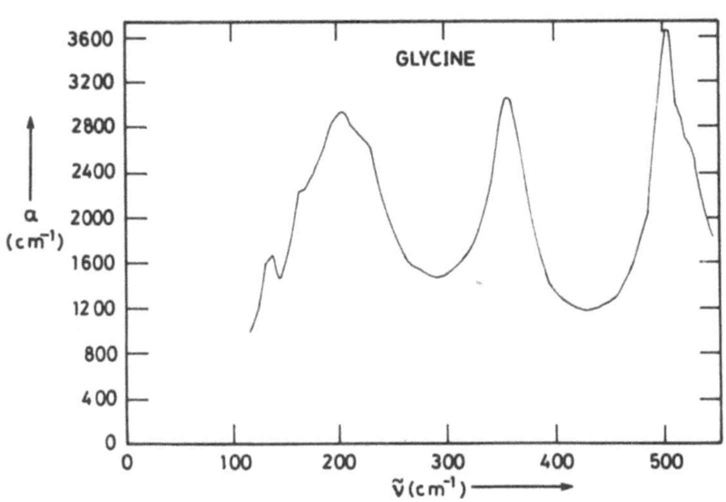

1d. Glycine

The spectrum of the polysaccharide cellulose (Figure 2a) displays sharper structure than is present in the protein spectra, and the bands are listed in Table 1, which includes also the bands for sucrose (Figure 2b) and glucose (Figure 2c).

Without deuteration and temperature variation measurements, at present in progress, confident assignments cannot be made.

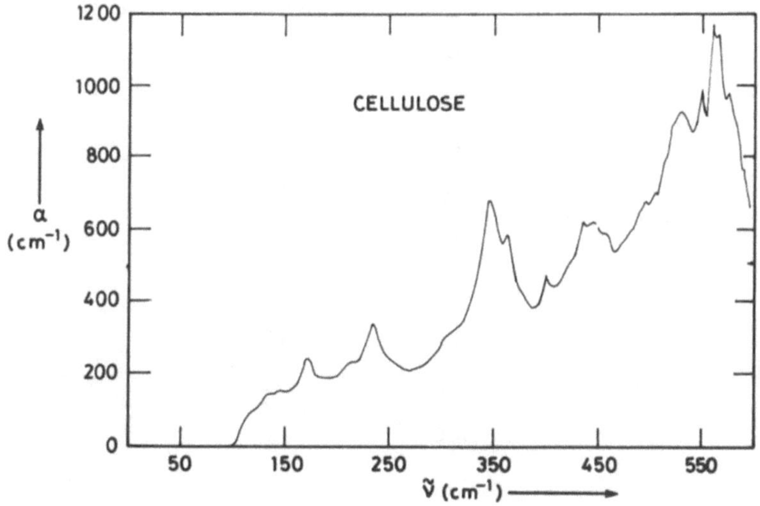

Figure 2. Absorption spectra of poly-, mono- and di-saccharides.
2a. Cellulose

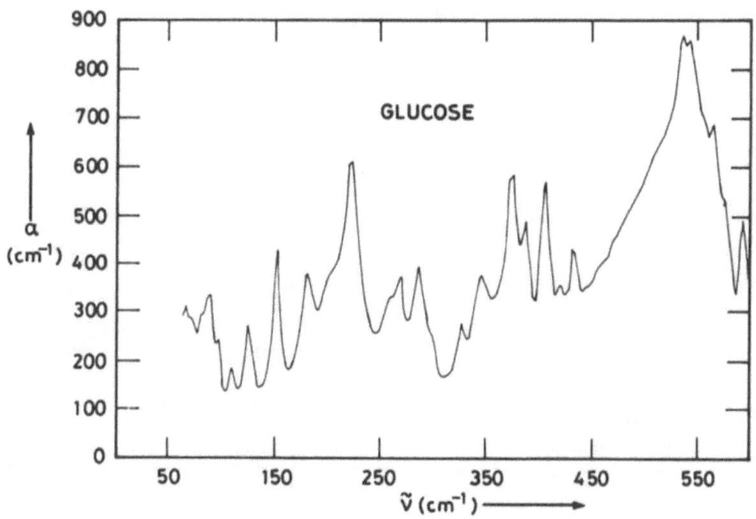

2b. Glucose

Table 1. Bands observed in aminoacids and saccharides (cm^{-1})

Glycine	Tyrosine	Sucrose	d-Glucose	Cellulose
90	32	49	48	75
106	70	61	70	101
138	90	75	89	172
170	114	86	98	213
205*	130	94	112	233
359	142	114	127	305
504	161	136	150	346
607	172	145	181	364
	180	159	208	373
	210	178	224	400
	250	195	261	425
	278	203	270	437
	310	214	287	447
	335	230	299	532
	380	235	329	565
	433	262	348	
	447	280	374	
	474	293	391	
	494	316	407	
	528	346	422	
	577	364	436	
		402	446	
		414	463	
		440	477	
		473	520	
		503	542†	
		530†		
		550†		

* structured

† possibly doublets

2c. Sucrose

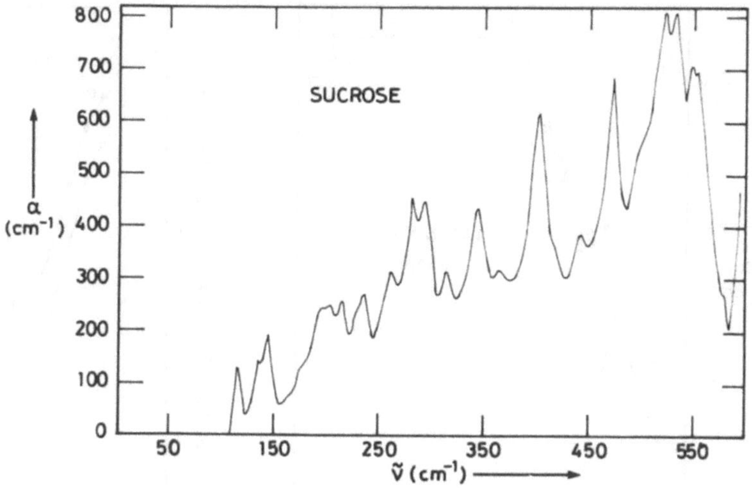

Low Frequency Dispersion Processes

For the study of these processes we arrange protein molecules in monomolecular layers using the Langmuir-Blodgett technique (2). A 2% solution of beef haemoglobin in 60% propanol, 40% 0.5 mol dm^{-3} sodium acetate (6) is allowed to spread on a clean water surface bounded on one side by a floating silk thread, and on another by a movable barrier. Study of the position of this thread enables the area A of the film to be measured. Suspension of a clean glass microscope coverslip, cutting the film and the water surface, from a chemical balance, enables the surface pressure of the film to be measured, so that movement of the barrier allows a pressure-area characteristic to be recorded, as in Figure 4. For the smallest surface pressures and largest areas (Figure 4, region a), the characteristic is curved, because the film does not cover the area completely; it is to be presumed that there are holes. At larger surface pressures and smaller areas there is a linear p(A) characteristic (Figure 4, region b) which is appropriate to a monomolecular film of known compressibility (7). This is the region in which the film must be maintained for extraction onto a substrate. As the area is decreased further by the movement of the barrier the film begins to collapse (region c), but at still smaller areas a further linear region (d) is found, possibly a bimolecular layer, followed in its turn by further collapse (e).

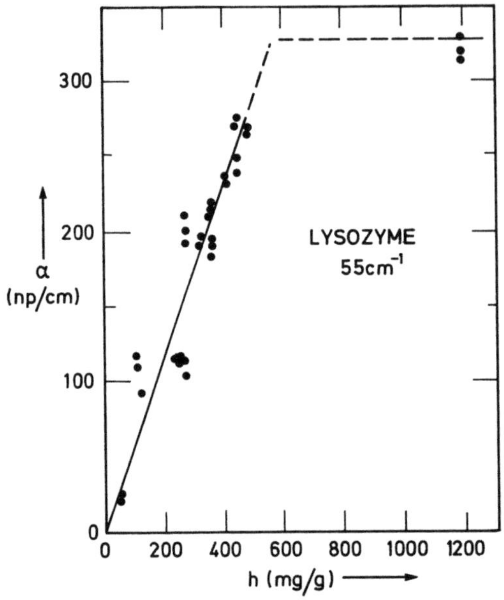

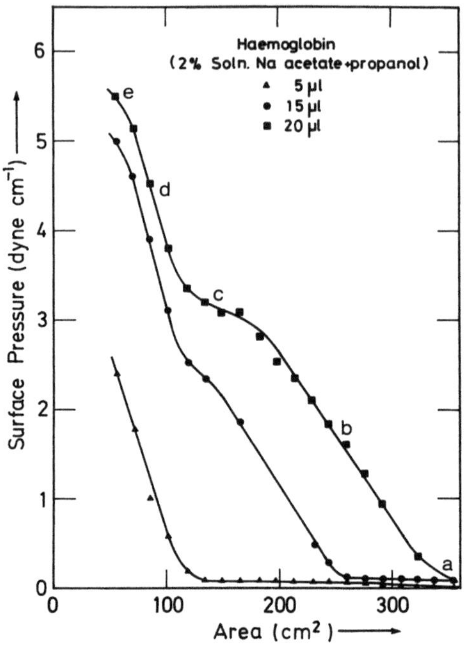

Figure 3. Variation with hydration of absorption at $\tilde{\nu}$ = 55 cm^{-1} of a cast film of lysozyme

Figure 4. Pressure-area curves for haemoglobin monolayers produced from 2% solution in propanol with sodium acetate. Triangles, 5 μℓ, circles, 15 μℓ, squares, 20 μℓ of solution applied to a trough of area 700 cm^2

A clean glass microscope coverslip, with evaporated aluminium electrodes, is mechanically lowered into and raised out of the water, at a uniform rate, each operation taking about fifteen minutes. The movement of the silk barrier is watched throughout and it is seen that as the glass is drawn out of the water, the film area is reduced, this reduction corresponding to the film area now deposited on the glass; but as the glass is lowered again, there is no corresponding area reduction, showing that for haemoglobin the next layer of film cannot be drawn down under the water. However, a second layer is deposited during the second draw-out. In the present experiments, the number of monomolecular layers corresponds only to the number of draw-outs; different conditions obtained in the previous experiments (2). Attempts to deposit the haemoglobin films on gold electrodes have not been successful.

Care must be taken to ensure that the films are adequately drained before the evaporation of a second aluminium electrode on their surface is attempted. Subsequently the films are allowed to stand in constant humidity (RH 55%, which is close to our normal laboratory humidity) for several days before electrical measurements are undertaken. Figure 5 shows the gradual attainment of time-independent capacitance and conductance by 4-layer films so treated. Further details of experimental technique are given in our first paper (2).

Inverse dependence of capacitance per unit area on the number n of layers is an important internal consistency check on these procedures. Figure 6 shows such a check, based on measurements in the flat 10 - 10^4 Hz region (Figure 7). Using a value ε = 20 appropriate to wet protein, we calculate, from the slope of the graph, the monolayer thickness to be 4.2 nm; from the intercept, the entire thickness of the aluminium oxide on the electrodes is calculated to be 3.8 nm. The fully hydrated haemoglobin dimensions are 5.3 × 5.3 × 7.1 nm (8).

Figure 8 displays the relaxation spectrum of 3- and 7-layer samples at the lowest available excitation voltage, 0.1 V. The data are essentially similar to those previously reported (2), but extended in frequency by the use of Wayne-Kerr and Marconi bridges. It is seen that there are two field-dependent relaxation processes. For 0.1 V across a 7-layer sample, which represents the lowest field (3.4×10^6 V/m) we have studied in the current series of experiments, the relaxations fall at 0.05 and 3×10^5 Hz. For the 3-layer sample the relaxations fall at 0.25 and 7×10^4 Hz.

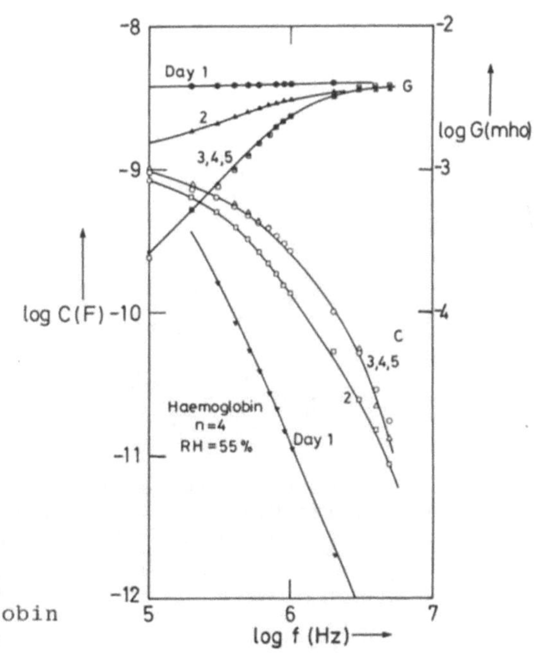

Figure 5. Variation of electrical characteristics of 4-layer haemoglobin film with period of humidification

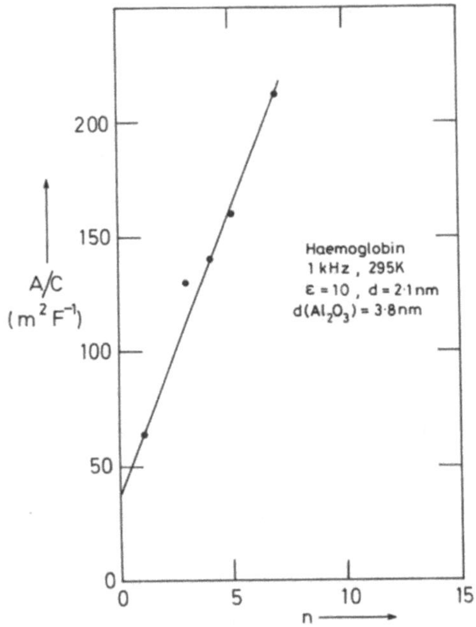

Figure 6. Variation of inverse of capacitance per unit area A/C with monolayer number n for haemoglobin at 1 kHz, 295 K

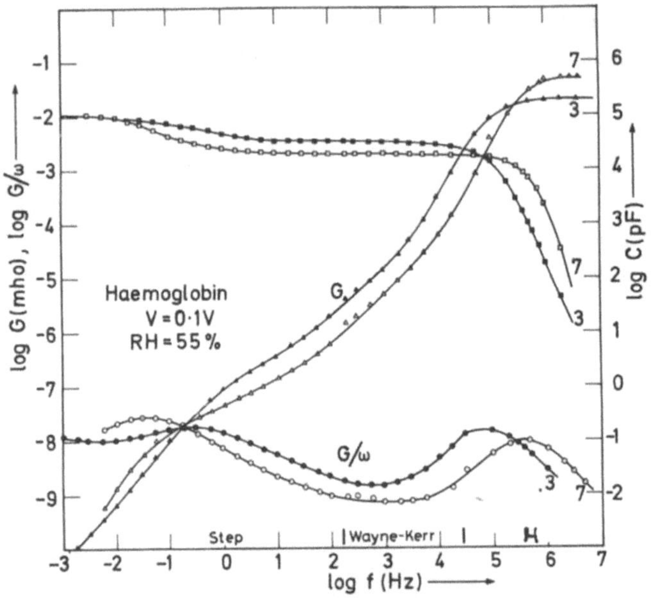

Figure 7. Electrical characteristics of 3- and 7-layer haemoglobin films for applied voltage 0.1 V, relative humidity 55%, as measured on step-function generator, Wayne-Kerr bridge and Hatfield bridge (H)

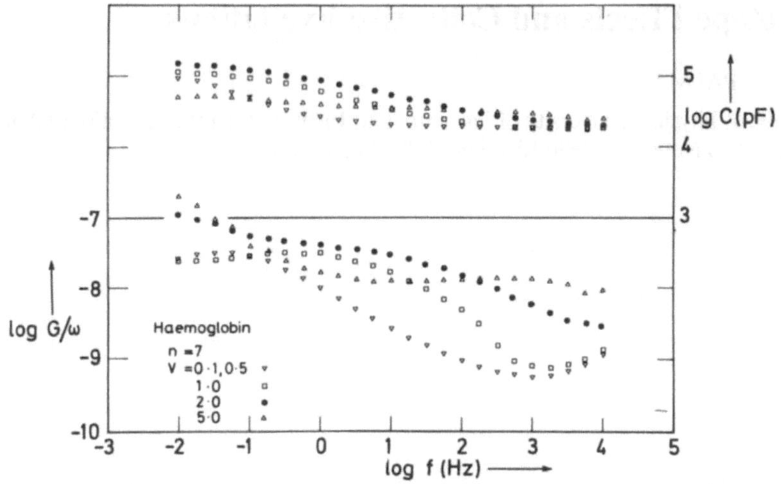

Figure 8. Field variation of electrical characteristics of 7-layer haemoglobin film

We have found using electrometry that a potential difference of about 15 mV exists permanently across a 7-layer film. This potential difference takes about τ = 4 days to re-establish after short circuit, following the relationship $V \propto 1 - \exp(-t/\tau)$.

It was found necessary that the samples be carefully drained by maintenance in the vertical position for a period of 5 hours before the electrode evaporation is carried out; in the first publication (2) this was not always realized. We have found it possible to reproduce the data of Figure 6 in that publication with insufficient draining; adequate draining results in reproducible data of the type of Figures 7, 8, below, in which the field dependence of the relaxations is maintained as before, although the capacitance is not anomalously enhanced.

The 7-layer sample has been measured with field strengths as high as 270 × 10^6 V/m; the data are shown in Figure 8. The high frequency relaxation is beyond the range of the measurements, but at the higher field strengths the low frequency relaxation process is to be found at sucessively higher frequencies, and is broadened:

Field (10^6 V/m)	Maximum loss factor frequency, for lower relaxation (Hz)
3.4, 17	0.05
34	0.5
68	2
170	500

Isotope Effects and Collective Excitations

M. U. PALMA

Dept. of Physics, University of Palermo, and Institute for Interdisciplinary Applications of Physics, Consiglio Nationale delle Ricerche, I-90123 Palermo

Summary

Isotopic substitution in the aqueous solvent is discussed as an effective method for probing the role of solvent dynamics in the stability of biomolecular conformation.

Introduction

A basic prerequisite to the microscopic understanding of biomolecular function is, as we know, the detailed knowledge of biomolecular geometric structure, and possibly of its response e.g. to changes of pH, of concentrations of other molecular species and electrolytes, of temperature, and so on. Despite the implicit restriction to only one half of the phase space (dynamic aspects being ignored), efforts in this direction were essential to the great historical advances of molecular biology and still maintain their significance in a variety of instances. In more recent times, the need of taking jointly into account both geometry and dynamics for a deeper understanding of biomolecular stability and function, and hopedly for the discovery of new phenomena (1), has been increasingly recognized. This is shown by the rapid growth of that part of current literature which deals with dynamic aspects of biomolecular physics, with their conjectured or already evidenced relevance to functional efficiency and stability, and more in general with the role of the time variable, that is with dynamics, fluxes and fluctuations (1-9). Indeed, it would appear arbi-

Coherent Excitations in Biological Systems
Ed. by H. Fröhlich and F. Kremer

References

(1) M.N. Afsar and J.B. Hasted, *J. Opt. Soc. Am.*, 67 (1977) 902.
(2) J.B. Hasted, H.M. Millany and D. Rosen, *J. Chem. Soc., Faraday Transactions 2*, 77 (1981) 2289.
(3) G.W. Chantry, H.M. Evans, J. Chamberlain and H.A. Gebbie, *Infrared Physics*, 9 (1969) 85.
(4) I. Golton, *Ph.D. Thesis*, Birkbeck College, University of London (1980).
(5) S.C. Shen, L. Santo and L. Genzel, *Can. J. Spectrosc.*, 26 (1981) 126.
(6) S. Ställberg and T. Teorell, *Trans. Faraday Soc.*, 35 (1939) 1413.
(7) J.T. Davies and E.K. Rideal, *Interfacial Phenomena*, Academic Press, New York, 1963.
(8) W.L. Bragg, E.R. Howells and M.F. Perutz, *Proc. Roy. Soc.*, A 222, (1954) 33.

Isotope Effects and Collective Excitations

M. U. PALMA
Dept. of Physics, University of Palermo, and Institute for Interdisciplinary Applications of Physics, Consiglio Nationale delle Ricerche, I-90123 Palermo

Summary

Isotopic substitution in the aqueous solvent is discussed as an effective method for probing the role of solvent dynamics in the stability of biomolecular conformation.

Introduction

A basic prerequisite to the microscopic understanding of biomolecular function is, as we know, the detailed knowledge of biomolecular geometric structure, and possibly of its response e.g. to changes of pH, of concentrations of other molecular species and electrolytes, of temperature, and so on. Despite the implicit restriction to only one half of the phase space (dynamic aspects being ignored), efforts in this direction were essential to the great historical advances of molecular biology and still maintain their significance in a variety of instances. In more recent times, the need of taking jointly into account both geometry and dynamics for a deeper understanding of biomolecular stability and function, and hopedly for the discovery of new phenomena (1), has been increasingly recognized. This is shown by the rapid growth of that part of current literature which deals with dynamic aspects of biomolecular physics, with their conjectured or already evidenced relevance to functional efficiency and stability, and more in general with the role of the time variable, that is with dynamics, fluxes and fluctuations (1-9). Indeed, it would appear arbi-

Coherent Excitations in Biological Systems
Ed. by H. Fröhlich and F. Kremer

trary and unphysical to assume that life does not make use of the wealth of possibilities conceivably offered by processes in the entire phase space and by spatial-temporal coherences.

A further extension is requested if we consider that biomolecule and solvent are two coupled subsystems of one functional system: their interaction, too, must be understood in terms which include both geometry and dynamics. Philosophically and historically, the interest of jointly considering geometry and dynamics of coupled subsystems is well recognized, since it had a basic role in the merging of crystallography and phenomenology of empirical material science into solid state physics: collective dynamics and dynamic interactions between subsystems such as conduction electrons or polarizable hydrogen ions, and crystal lattice had to be deeply understood before notions such as superconductivity, polarons, ferroelectricity, and many more, could be sorted out and mastered.

Deuteration affords a most convenient modulation of the mass, that is of a dynamic parameter, in the aqueous subsystem. The inertial changes due to D-H substitution are between 2:1 and 20:18 in the solvent, depending upon the type of motion involved, and their effect can be detected. Corresponding inertial changes in solute biomolecules are negligibly small, usually, since deuteration of exchangeable hydrogen sites does not cause appreciable overall mass alterations. It should be remembered, however, that solvent deuteration can alter the strength of hydrogen bonds involving hydrogens at exchangeable sites in biomolecules. Further, the slight difference between Deuterium- and Hydrogen bond lengths might involve additional, if slight, deformation energies in solute biomolecules. Effects of this type must be carefully discriminated, and indeed they can (10-13).

Two Interacting Subsystems

Let us briefly recall a simplified picture of two interacting subsystems in a mean field approximation (14). Let us refer to the diagram in Fig. 1, as adapted from Ref.13. Here the macroscopic state of the system is described by the two order parameters S_1, and S_2 of subsystems 1 and 2 respectively. The equations of state of the two isolated subsystems are of the type (14):

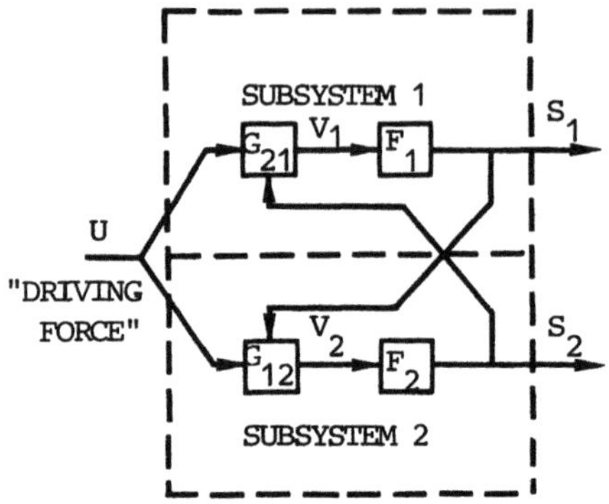

Figure 1. a system composed of two interacting subsystems. (Adapted from Ref.14)

(1) $$S_i = F_i(V_i)$$

where V_i is the driving force acting upon the subsystem i, expressed as an effective field resulting from the combined action of the external field U ("driving force") and of S_j:

(2) $$V_i = G_{ji}\ (U,S_j)$$

the G_{ji}'s being coupling functions. The driving forces V_i themselves are written in a linear approximation as:

(3) $$V_i = U + K_{ji}S_j$$

Now, if at least one of the subsystems exhibits phase transition(s), a variety of stable solutions are available for the whole system. In Fig. 2 (again adapted from Ref.14) we show the simplified case obtained under the assumption that one of the two subsystems exhibits a first-order phase transition, and the other just behaves monotonically. The figure shows on its right side many regions of the (K_{ij}, K_{ji}) plane, corresponding to different behaviours. On the left side are shown behaviours corresponding to regions R and H. Those predicted for regions N and I are not shown: they correspond, respectively, to quasi-decoupling and to instabilities and oscillations (14).

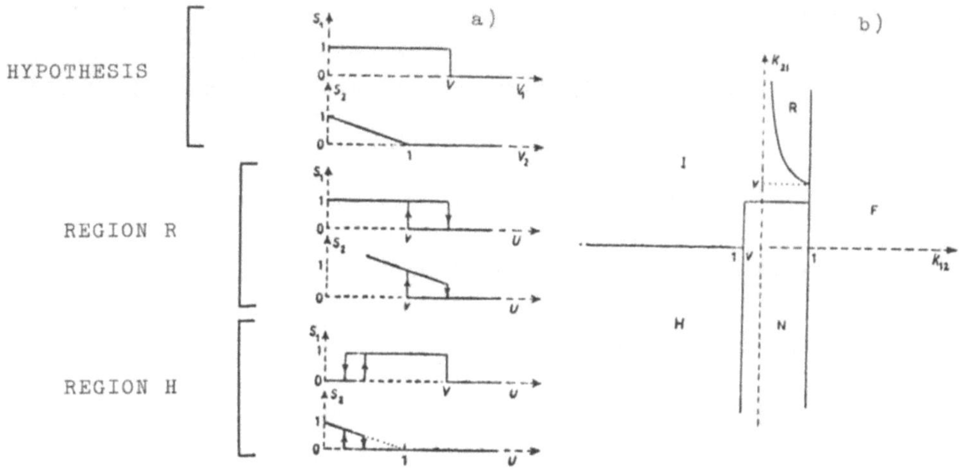

Figure 2. a) an example of different stable solutions for the system of Fig. 1, under a most simple hypothesis.
b) corresponding regions in the (K_{12}, K_{21}) plane (adapted from Ref.14)

Although oversimplified, this scheme helps clarifying the use of solvent deuteration for detecting a possible role of solvent dynamics (as affected by solute-solvent interaction) in the conformational/functional stability of biomolecules. We may: i) select a biomolecular species showing conformational transition(s) in aqueous solution ($=S_1$); ii) substitute D_2O for H_2O, that is: change the state equation $S_2 \equiv F_2(V_2)$ and/or the coupling K_{21}); iii) if transition(s) are affected, the K_{21} coupling is non-zero: interpret.

Clearly, this has not much in common with "solvent isotope effects" (15), (16).

Solvent Deuteration And Biomolecular Stability

We start discussing the stability of helical structures obtained by mixing aqueous solutions of two synthetic polyribonucleotides: polyriboadenylic acid (poly A) and polyribouridylic acid (poly U), and we shall refer to Ref. 12 for results and further references. Poly(A) and poly(U) are known to provide useful model systems for the study of conformational stability of nucleic acids. They can form two well defined helical complexes: the double stranded poly(A). poly(U) with a 1:1 monomer ratio, and the triple stranded poly(A).2 poly(U) with a 1:2 monomer ratio. Four conformational transformations are possible, i.e.:

(4) poly(A).2 poly(U) = poly(A).poly(U) + poly(U)

(5) poly(A).poly(U) = poly(A) + poly(U)

(6) poly(A).2 poly(U) = poly(A) + 2 poly(U)

(7) 2 poly(A).poly(U) = poly(A).2 poly(U) + poly (A)

From left to right, these are observed as sharp, cooperative transitions. It has been found that the corresponding stability diagram is affected by solvent deuteration as shown in Fig. 3 (adapted from Ref. 12 with some simplifications).

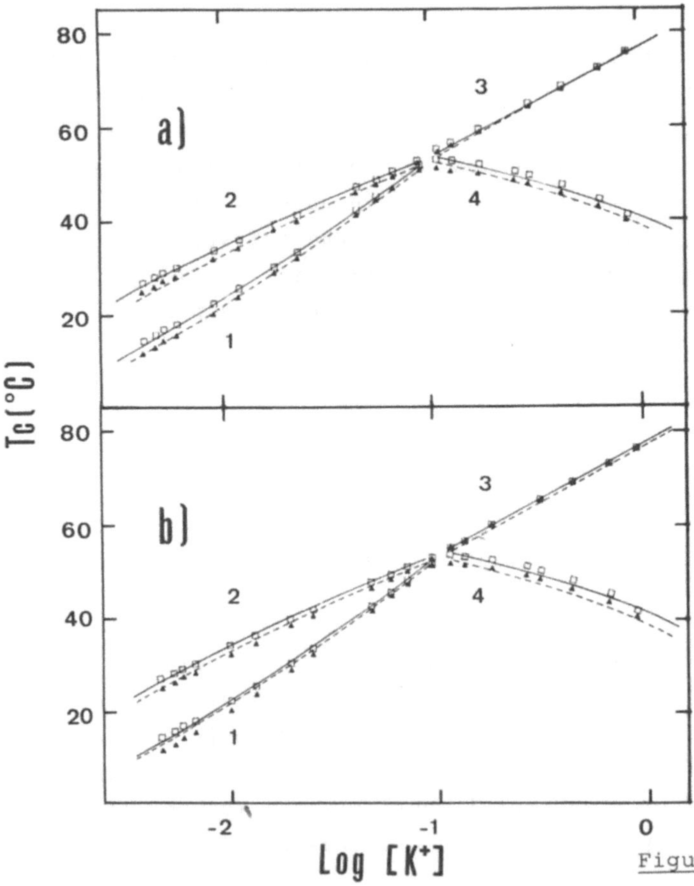

Figure 3. (Legend see p. 89)

Fractional deuteration causes fractional shifts of critical temperatures, leaving (except for some special cases discussed in Ref. 12) the transition profile unchanged, as expected on the basis of the cooperative character of the transitions (12). Precise computer fittings of experimental data, based on thermodynamic analysis have also been performed (12). In Fig. 3 we see two such fittings. In the case of Fig. 3a), all parameters were left adjustable in the course of optimization, whose final result indicated that the isotopic modulation of the solvent caused a modulation of the entropy but not of the enthalpy term at helix - coil transitions. To check this conclusion independently, in another optimization the entnalpy parameter alone was left free to be adjusted. The best fit so obtained is shown in Fig. 3b). Altogether, this shows that solvent deuteration stabilizes these helical structures and that the stabilizing effect should be ascribed to an isotopic modulation of the entropy term at helix-coil transitions, enthalpies remaining essentially unaffected (12).

Before discussing the sign of this entropy modulation, we shall briefly review other cases, all pointing to the same conclusion. Native DNA is far more complex than poly(A).poly(U) but it, too, appears to be stabilized by solvent deuteration. For B. subtilis DNA it has in fact been shown by X-ray microscopy (17) that at temperatures close to the helix-coil transition, the profile of the percental "open loop" regions vs. temperature is isotopically shifted much as in the case of branch 2 in Fig. 3. These early, visual data will appear in the discussion at the end of the present paper to have further relevance, in view of the relation between solvent dynamics and nucleic acid dynamics (3).

A much simpler, yet clarifying case is that of Metilene Blue, a small planar molecule tending to aggregate in stacked arrays and for this reason a useful model system for the study of stacking base pairs in nucleic acids. In this case, it has again been found that solvent deuteration causes the same kind of modulation in the entropy term of the stacking process (13).

Figure 3. example of the effect of solvent deuteration on nucleic acid ordered structures. Branches (1) to (4) correspond respectively to eqs. (4) to (7) in the text. Full triangles show experimental data in the H_2O case and are computer fitted with full lines. Figures a) and b) differ in the way fittings were obtained. (Adapted with simplifications from Ref. 12).

In the case of biomolecules possessing charged groups, the isotopic modulation of entropy might be due in part to the trivial expected change of dieletric constant (12). This is one of the reasons making the case of Agarose interesting, an uncharged polysaccaride showing thermoreversible helix-coil and supramolecular ordering transitions (18). Solvent deuteration affects both transitions. A van't Hoff analysis has shown that the more ordered phase is stabilized by an isotopic alteration of the entropy term, similar in its magnitude and sign to those that we have just discussed (11), (19). It should be remarked that, although van't Hoff analysis should be used with great caution in this type of systems (11), the rather small size of the observed effect legitimates "ceteris paribus" comparative conclusions to be drawn (11). Further, solvent deuteration causes a similar entropic stabilization of supramolecular structure (19) and, accordingly, it also affects the delay time observed in the onset of a hitherto unobserved type of gelation, giving rise to very long range order (19).

In all cases discussed so far, solvent deuteration stabilizes ordered configurations of solutes (or equivalently, destabilizes disordered configurations). Unavoidable intramolecular deuteration at sites where exchange is allowed, and the consequent presence of (stronger) intramolecular deuterium bond seems to be compensated or to play no role: thermodynamics shows that in all cases studied the effect is essentially entropic (11), (12), (13). Comparison of the charged and uncharged cases shows that a role of the higher dielectric constant of D_2O would be very unlikely (11), (12), (19). We are thus led to conclude that deuteration <u>lowers</u> the entropy of the entire system when the solute is in the disordered configuration (11), (12), (13). The stabilizing effect is small (1-3 ^{o}C) but, as we are going to see, significant.

Discussion

An ample literature is available on the existence and the effects of biomolecule-solvent interactions (20-30). We shall now see how the small isotope effects observed can illustrate new aspects of these interactions which concern the role of water dynamics in biopolymer conformation. We consider the Gibbs free energy associated with a conformational transformation of a solute:

(8) $$\Delta G = \Delta G_1 + \Delta G_2 + \Delta G_{12}$$

where subscripts 1,2, and 12 stand for "solvent", "solute", and "solute-solvent" respectively. Now, ΔG_1 is known to be zero (27), so that it must remain unaffected by deuteration. The remaining contributions will be conveniently distinguished in their entropy and enthalpy components. Since all results point out that deuteration hardly affects enthalpies, either ΔH_2 and ΔH_{12} are not affected, or they are affected by compensating amounts. The wanted entropic contribution could very unlikely come from ΔS_2 as affected by the lower dielectric constant of D_2O, in view of the discussed independence of the results upon the presence of charged groups in the biopolymer. It must therefore come from ΔS_{12} (11), (12), (13). The mere difference between hydrogen and deuterium bond energies cannot explain the experimental results.

This conclusion can be viewed in terms of the higher strength of hydrophobic interactions in D_2O, and of the hiding of the biopolymer hydrophobic groups away from the solvent in the polymer ordered configuration (11-13), (20-31). A better understanding is possible, however, by remarking that a mere change of mass -a dynamic parameter in the solvent, entrains and shifts a conformational transition in the biomolecule subsystem. This evidences the existence of a coupling (K_{21} in Fig. 2) between the dynamics of one subsystem (the solvent), and the conformation of the other (the solute biopolymer).

Current notions on both bulk and hydration water (30-34) offer more insight in a possible microscopic mechanism. Liquid water should be viewed as a flickering network of uninterrupted hydrogen bond paths, spreading in all directions and connecting the liquid in the large (30). Closed paths are also expected to occur (30). Along the network of "connectivity pathways" there will be a non-vanishing order parameter, implying the existence of a local electric field. The order will be propagated along the pathway, by way of a geometric-dynamic feedback, whereby the local field will limit the dynamic orientational freedom of single water molecules, and this in turn will prevent complete washing-out of the field at the next point of the connectivity network (3), (30). Deuteration will alter a dynamic parameter relevant to this mechanism, and thus it is expected to alter the geometric-dynamic structure of liquid water (2).

Across a conformational transition, a solute biopolymer will alter the proportions of its hydrophilic and hydrophobic areas exposed to the solvent. The former are expected to favour the "imprinting" of connectivity pathways, which will have a lifetime one or two orders of magnitude longer than in bulk water (3), (31). Monte Carlo calculations provide spectacular evidence for the existence of such "pinned connectivity pathways" in both DNA and Agarose structures (32-34). Hydrophobic surfaces instead, will force the network of connectivity pathways to rearrange itself in a convex cage, with a resulting negative entropic contribution due in part to the strengthening of the bonds, and in part to the reduced number of independent ways in which the network can be constructed (30). As a result, the structure of the solute biopolymer and that of the connectivity networks allowed in the solvent will depend on each other. The latter, in turn, depends upon dynamic parameters of the solvent molecules. This illustrates how solvent dynamics (collective, and thermally excited) can be directly involved in the structural stability of biomolecules, and it offers an interpretation of this stability in a far ampler volume of the phase space. It should be remarked that solvent dynamics in the large enters the picture, inasmuch as it is probed and averaged out by the flickering connectivity network. We may wonder if this could offer a basis for the microscopic interpretation of a most interesting variety of results obtained by solvent perturbation by cosolvents (25), (26).

Conclusions

Deuteration of aqueous solvent tends to stabilize solute biopolymers in their ordered conformation. The effect is small and it may appear trivial, for it has long been known that deuterium increases the strength of hydrogen bonds and of hydrophobic interactions. As it turns out, however, stabilization is due not only (if not at all) to the mere enthalpic effect due to the increased strength of hydrogen bonds, or to the decreased dielectric constant. There must be an entropic contribution, and this must derive from solute-solvent coupling. The dependence upon the mass of hydrogen makes it clear that this coupling occurs between the geometry of the solute and the dynamics of the solvent, and that differences imposed to the solvent dynamics by a conformational transition of a solute biomolecule, effectively concur in providing and stabilizing the ordered, and functional, biomolecular shape. This, united to the current interest in the dynamics of biomolecules themselves, allows viewing in the entire phase space the problem of the stability of biomolecules in their solvent, as it is indeed philosophically desirable. Current views on the structure of liquid water provide a basis for speculations on a possible class of microscopic models. The specific model discussed in this paper should be viewed just as a help to visualize the type of processes involved.

I wish to express my sincere thanks for many clarifying discussions to all colleagues whose work I have quoted, and particulary to Prof. M.B. Palma-Vittorelli; and to Ms. M. Genova-Baiamonte and Ms. V. Paladino for producing the processed typescript.

References

(1) Fröhlich, H., in "From Theoretical Physics to Biology" (M. Marois, Ed.), North-Holland Publishing Co., Amsterdam, (1969), and Adv. Electronics and Electr. Phys. 53,85 (1982).
(2) Aiello, G., Micciancio-Giammarinaro, M.S., Palma-Vittorelli, and Palma, M.U., in "Cooperative Phenomena" (Haken, H. and Wagner, M. Eds.), Springer Verlag, Berlin, (1973).
(3) Palma, M.U., in "Structure and Dynamics: Nucleic Acids and Proteins" (Clementi, E. and Sarma, R.H. Eds). Adenine press, (1983), p. 125.
(4) Careri G., in "Quantum Statistical Mechanics in the Natural Sciences" (Kursunoglu, B., Mintz, S. L., and Widmayer, S. M.), Plenum Press, N. Y., (1974).
(5) Austin, R.H., Beeson, K.W., Eisenstein, L., Frauenfelder, H., and Gunsalus, I.C.: Biochemistry, 14, 3555 (1975), and Frauenfelder, H., in "Mobility and Function in Proteins and Nucleic Acids" (CIBA Foundation Symposium 93), Pitmann, London, 1982.
(6) Davydoff, A.S., "Biology and Quantum Mechanics", Pergamon Press Oxford (1982).
(7) Englander, S.W., Kallenbach, N.R., Heeger, A.J., Krumhansl, J.A., and Litwin, S.: Proc. Natl. Acad. Sci., (USA) 77,7222 (1980).
(8) Scott, A.C., Phys. Rev. A, 26, 578 (1982)
(9) see, e.g. Karplus, M., and McCammon: Nature 277, 578 (1979).
(10) Vento, G., Palma, M.U., and Indovina, P.L., J. Chem Phys., 70, 2848 (1979).
(11) Indovina, P.L., Tettamanti, E., Micciancio-Giammarinaro M.S., and Palma, M.U.: J. Chem. Phys, 70 1'-2', 2841 (1979).
(12) Cupane, A., Vitrano, E., San Biagio, P.L., Madonia, F., and Palma, M.U.: Nucl. Acid Res., 8, 4283 (1980).
(13) Fornili, S.L., Sgroi, G., Izzo, V.: J. Chem. Soc., Faraday Trans. I 1983, 79. (in press).
(14) Micciancio, S. and Vassallo, G.: N. Cim. (D), 1, 627 (1982)
(15) See e. g. Collins, C. J. and Bowman, M. S. (Eds): "Isotope Effects in Chemical Reactions" - Van Nostrand Reinhold Co., N. Y. 1970, and list of references there.
(16) Halevi, E. H.: Israel J. Chem. 9,385 (1971).
(17) Fornili, S.L. et al.: Nucleic Acid Res. 2, 1805 (1975).
(18) Arnott, S., Fulmer, A., Scott, W.E., Dea, I.C.M., Moorhouse, R., and Rees, D.E.: J. Mol. Biol. 90, 269 (1974).
(19) Leone, M., Fornili, S.L., Palma-Vittorelli, M.B., and Migliore, M.: 2 nd Int'l Conf. on Water and Ions in Biologic Systems (Bucharest, 1982), and work in progress.
(20) Kauzmann, W.: Adv. Protein Chem. 14 1 (1959).

(21) Franks, F. (Edit.): "Water, a comprehensive Treatise", Plenum Press (seven volumes) (1972-1982).
(22) a) Scruggs, R.L., Achter, E.K., and Ross, P.D.: J. Mol. Biol. 47, 29 (1972).
b) Porschke, D., and Eggers, F.: Eur. J. Biochem. 26, 490 (1972).
c) Alvarez, J., and Biltonem, R.: Biopolym. 12, 1815 (1973).
(23) Bixon, M., and Lifson, S.: Biopolym. 4, 815 (1966).
(24) Edelhoch, H., and Osborne, J.C. jr: Adv. Prot. Chem. 30,183, 1976.
(25) Cordone, L., Cupane, A., San Biagio, P.L., and Vitrano, E.: Biopolym. 20, 53 (1981).
(26) Cordone, L., Cupane, a., D'Alia F., and De Stefano, M.G.: Far. Symp. Chem. Soc. 17 (in the press). and list of references there.
(27) Ben-Naim, A.: Hydrophobic Interactions (Plenum Press) N.Y. (1980.)
(28) Kreshek, G.C., Schneider, H., and Scherage, H.A.: J. Phys. Chem. 69 3131 (1965).
(29) Ben-Naim, A., and Wilf, J.: J. Chem. Phys. 70, 771 (1979).
(30) Stillinger, F.H.: Science 209, 451 (1980).
(31) Rowland, S.P. (Editor): "Water in Polymers" A.C.S. Symp. Series no. 127 (Am. Chem. Soc., Washington, D.C. 1980).
(32) Clementi, E., and Corongiu, G.: Int'l J. of Quan. Chem. XXII, 595 (1982).
(33) Clementi, E.: "Structure and Dynamics: Nucleic Acids and Proteins" (Clementi, E. and Sarma, R.H. Eds.). Adenine press, (1983), p. 321.
(34) Corongiu, G., Fornili, S., and Clementi, E.: work in progress (private comm.).

Long-range Energy Continua in the Living Cell: Protochemical Considerations

G. RICKEY WELCH[1] and MICHAEL N. BERRY[2]

[1] Department of Biological Sciences, University of New Orleans, New Orleans, LA 70148, USA
[2] Department of Clinical Biochemistry, Flinders University, Bedford Park, SA 5042, Australia

"The visible world is neither matter nor spirit, but the invisible organization of energy" /1/.

I. INTRODUCTION

When we approach life at the level of the single cell and enter the domain of cellular biochemistry/biophysics, we find (contrary to the picture given in standard textbooks) a highly fragmented status rerum, largely dominated by an air of reductionism (viz., 1+1 = 2). Most attempts at understanding the essence of the "living state" at this level have a materialistic basis. Such an outlook all too often fails to relate the fact, that living systems are, by their nature, defined in a dynamic sense. Hence, we should study the cell by "reducing" it, not to its elements of matter, rather to its elementary processes. A "process" view imparts an emphasis on energetics and energy flow. Here, we are obliged to apply a lesson gleaned from physics - that of matter/energy duality in physical systems. One finds, that matter is the substance of things, while energy is the moving principle /2/.

When we regard energy and cell metabolism, we usually think of ATP in its role as a "molecular", diffusible carrier of chemical free energy. Despite the ubiquity (and universality) of ATP as a free-energy "currency", there remain fundamental questions concerning energy and metabolism. How much "individuality" is there in the processes of cell metabolism? Is the metabolic structure of the cell a jig-saw puzzle or an interlocking web? How accurate is our quantitative assessment of the kinetic and energetic aspects of metabolic processes, based on the study of counterpart processes in vitro? Are there long-range modes

Coherent Excitations in Biological Systems
Ed. by H. Fröhlich and F. Kremer

(continua) of free-energy transduction extant in vivo? - Such questions present themselves as bona fide elements in a new era of synthesis in biochemistry and biophysics.

Generally, any kind of holistic conception, as to the existence of energy continua in the living cell, has not been deemed palatable by the sciences of biochemistry and enzymology - in their traditional, reductionistic posture. We contend, that the reason for rejection of such ideas has been the lack of appreciation, until more recently, of the following specific factors: i) the reactivity of the very protein fabric of the enzyme molecule, ii) the material substratum in which most metabolic processes are imbedded in vivo, and iii) the essence of motive energy-transducing principles. As propounded most notably by Lumry and coworkers /3,4/, regarding factor i), the protein and chemical subsystems must engage in a fluid and variable exchange of free energy during the course of enzyme catalysis. Concerning factor ii), it now appears that the majority (perhaps all) of the enzymes of intermediary metabolism exist in vivo in organized states, whose "sociological structure" is far different from that of the isolated enzyme molecules in vitro /5,6/. Permeating subcellular particulates, to which many of these enzyme systems are adsorbed, are strong (local) electric fields and mobile protonic (and, possibly, electronic) states. As regards factor iii), these "external" devices may entail unique modes of free-energy transduction, as well as facilitation and coordination of enzymatic processes.

In the present treatment, we wish to focus on the mobile protonic mode ("proticity") as a motive-transducing principle. An electrochemical interpretation of cell metabolism, developed recently by Berry /7/, provides the basis of our study. Here, we will extend this interpretation and examine the basic properties, of the protein molecule and of organized multienzyme systems, which suggest roles of proton flow in the integration of cell metabolism.

II. THE STRUCTURAL NATURE OF METABOLIC PROCESSES

A. A Biphasic Picture of Cellular Infrastructure

In recent decades electron microscopy has revealed a complex and richly diverse particulate infrastructure in living cells - especially larger eucaryotic cells. This structure encompasses the extensive membraneous reticulation (e.g., plasmalemma, endoplasmic reticulum, mesosomes [bacteria]), as well as the hyaloplasmic space. The latter region, containing the so-called "ground substance", is laced with a dense array of various cytoskeletal elements (e.g., microtubules, microfilaments, intermediate filaments) and an interlocking micro-trabecular lattice. (See Prof. Clegg, this volume).

Accumulating empirical and theoretical considerations indicate that the majority (if not all) of the enzymes of intermediary metabolism operate in vivo in association with particulate structures. A perusal of the literature reveals evidence for enzyme organization in virtually all major metabolic pathways. The organizational mode, of the multienzyme systems of specific metabolic processes, may entail formation of protein-protein complexes and/or individual adsorption to cytological stubstructures. Strong evidence has come from centrifugation studies on whole cells /8,9/ and cell fragments (10). Experimental and theoretical calculations /11,12/ of protein "concentrations", in association with cytomembranes and organelles, indicate high, crystal-like densities of protein molecules in (on) particulate structures of the cell. There is a remarkable homology, in the surface area-to-volume ratio for all membraneous cytological substructures. Such considerations led Sitte /11/ to propose, that all cytomembraneous elements have evolved in a common fashion to function as effective "protein collectors" in the operation of cell metabolism. And, recent work from Porter's group /13/ shows the microtrabecular lattice itself to be "dressed" with a multitude of (as yet unidentified) proteins.

Thus, cell biology presents us with a rather simple, biphasic view of cellular infrastructure: a solid phase, emcompassing extensive membrane surfaces and the fibrous lattice-work; and a soluble, aqueous phase (albeit containing a considerable amount of structured water [see Prof. Clegg, this volume]). We must focus on the solid phase as the primary site of intermediary metabolic processes, with the soluble phase functioning largely in such subservient roles as thermal buffering, distribution of common substrates, regulatory substances, and salt ions, etc.

B. The Logic of Enzyme Organization

Reaction-Diffusion Flow in Organized States

We find the study of biochemistry at a point in its development, prophesied accurately some 50 years ago by Peters /14/, whereby "in the ultimate structural units of the cell, we have reached the limit for the application of ordinary statistical considerations, and must substitute some more anatomical view, based upon control by surface... Owing to the microheterogeneous nature of the system, surface effects take precedence over ordinary statistical, mass action relationships and become in the ultimate limit responsible for the integration of the whole and therefore the direction of activities." Within the confines of the ultimate "metabolic microenvironments" of the living cell - engendered by the afore mentioned organizational modes, along with the ambience of structured water, electrical double layers, etc., at the surface boundaries - we must abandon the concept of scalar chemical reactions, as well as such familiar notions as that of a "uniform concentration." The simplicity, homogeneity, and isotropy of the in vitro condition are artificial and deceptive. Traditionally, we have defined "enzymes" simply as "proteinaceous catalysts". Now, we must add more biological flavor, asserting that enzymes catalyze specific reactions at rates and under appropriate conditions, commensurate with the vitality of the cell - with the idea of "locational specifity" /15/ part-and-parcel of the defining character /6/.

In these localized microenvironments, traditional "macroscopic" descriptions of metabolic processes, employing the standard differential equations for reaction/diffusion dynamics, simply break down /5/. In particular, the idea of a "bulk concentration" will not apply, in most instances, in the cellular microenvironments. Here, concentrations of enzymes and their respective substrates are, in many cases, of the same order of magnitude. It is highly plausible, that there are "molecular channels" in organized multienzyme systems in vivo, wherein each individual enzyme is subject to a local, "quantized" substrate concentration /5/. A similar "channel" picture (the so-called "mosaic chemiosmosis" model) has been proposed for the coupling of proton pumps to H^+-ATPase (see Dr. Kell, this volume).

Similary, the "macroscopic" approach is inadequate for depiction of the overall, vectorialized material flow in reaction-diffusion systems exhibiting the kind of inhomogeneity and anisotropy as those in vivo. We need a physicomathematical construct that will relate the manner in which the physical components are "hooked-up". The most promising attempt, to date, is embodied in the network theory /5/. Such a mathematical formalism goes a long way toward describing how things are "hooked-up", both structurally and functionally, in the intermediary metabolism of the living cell. Yet, something is inherently missing from the picture; some physical (nay, biophysical) "integrative principle" is lacking. Just how are the elementary processes, in an organized multienzyme sequence in vivo, coordinated to produce a "sociological unit" /6/? From holistic inclinations, we intuit that there must be integrative factors which "zip" an enzyme system into a whole, which subordinate the individual events (energetically) to the good of the whole. In short, we need some indication of how metabolism works in the organized state, in addition to our knowledge of how the components are arranged physically.

Traditionally, we have tended to characterize, thermodynamically, the "organization" of coupled reaction-diffusion processes in terms of the free-energy flow therein. This course has taken us far, in our attempts to understand the nature (and chemical "directionality") of biological processes. However, elucidation of energetic principles applicable to the microscopic confines of organized states in vivo, demands that the usual thermodynamic analysis be supplemented with molecular details of the functional coupling /16/. Without such information, one is limited to a "black box" approach - able just to define global parameters.

We suggest, that one "integrative principle", potentially widely applicable to material flow in organized enzyme systems, was proposed 25 years ago - a principle which has heretofore seen rather restricted (though increasing) usage. The very inception of the concept of "vectorial metabolism" dates to the pioneering work of Mitchell /17/, in the late 1950's, dealing with group-translocation and chemiosmotic reaction systems. He realized, early on, that chemical catalysis by porters, may be regarded as occurring by specific ligand-conduction mechanisms. That is to say, the chemical reaction taking place within the active center of an enzyme is, by the virtue of inherent anisotropy in the arrangement of reactive groups, a local vectorial process. And, Mitchell recognized that, by plugging enzymes through membranes or organizing them in complexes, this local effect would generate a macroscopic vectorial flow.

A basic motive in Mitchell's work has been the elucidation of general mechanisms of linkage between metabolism and transport. This "vectorial idea" found one of its earliest, and most notable, applications in the bioenergetics of electron-transport phosphorylation. Most importantly, this work spawned the concept of "proticity" and the role of the proton gradient as a bioenergetic transducing principle. The "proton gradient" has emerged as a central, interconvertible form of energy - interconnecting such diverse entities as electron transport, ATP-bond energy, transhydrogenation reactions, flagellar (mechanical) motion, pH regulation, active transport, and heat production. When the cell must ensure flow of a substance in a preferred direction across the organized phase of a membrane, transport is coupled to a local, redox proton-motive-force generator or to an ion-translocating ATPase (which, itself, may serve as a proton-injector /18/).

We would like to extend further the biological role of proticity - as an integrative force in the material transformations of intermediary metabolism. Organization of enzyme sequences at cytosol - particulate interfaces juxtaposes them to sources of proton gradients and significant electric fields. Is this juxtaposition fortuitous? Are the localized enzymatic processes in vivo functionally independent of those energy-transducing elements? We suggest not. In the next two sections, we will suggest that the "reactivity" of the protein matrix, in conjunction with the substratum of the afore mentioned organizational modes, may be "designed" (teleonomically) to make full use of such energy sources. As we shall see, this functional coupling potentially entails long-range energy continua which add a degree of coherency to cell metabolism.

III. ENZYMES: BIOCHEMICAL ELECTRODES AND PROTODES

Let us consider the "reactive" properties of the protein molecule, which suggest that enzymes are suited to function as field-effect electronic/protonic elements in the execution of chemical reactions.

A. Structural-Functional Bases of Enzyme Action

Throughout the 100-year history of enzymology /19/, the major emphasis has been on the active center of the enzyme molecule. We have used the transition-state free-energy change, $\Delta G^{\ddagger}$, as a "window" into the workings of this reaction site, decomposing catalytic processes into entropic ($\Delta S^{\ddagger}$) and enthalpic ($\Delta H^{\ddagger}$) contributions. Indeed, all of the customary sources of catalytic power can, via $\Delta S^{\ddagger}$ or $\Delta H^{\ddagger}$, be formulated in the guise of transition-state stabilization /20/.

From the enthalpic ($\Delta H^{\ddagger}$) side, one finds in the enzymology literature a number of catalytic (energetic) factors which lead to decrease in $\Delta G^{\ddagger}$. Generally, these fall into one of the three categories: i) catalysis by "rack" mechanisms (whereby the bound, ground-state substrate is strained or distorted), ii) general acid-base catalysis (involving proton-transfer to or from the transition state), and iii) electrophilic/nucleophilic catalysis (usually involving covalent stabilization of the transition state). Underlying these various catalytic modes is the importance of local electric fields, in the microenvironments of enzyme active centers. It has been suggested, that the geometry of active-center dipole groups is engineered in evolution to provide optimal "solvation" (electrostatic stabilization) of polar transition-states /21/.

What of the role(s) of the protein matrix in active-center events? Considering the relatively small volume occupied by the actual reaction site, one might ask, why are enzymes so big? It is now apparent, that the protein as a whole is designed to provide a specific solvent medium for a given chemical reaction /22/. Wherein, the combined chemical and protein subsystems engage in a fluid and variable exchange of free energy, facilitating the entrance of the bound chemical system into its transition state. Accordingly, we are led to picture the protein matrix as an intermediary, a "deterministic" mediator, between a localized chemical reaction-coordinate and the ambient medium. A crucial point here is that, while the local active-site configuration defines the physicochemical nature of the substrate-product transition state(s), the protein molecule determines the rate at which this state(s) is reached.

Viewing the enzyme as a macromolecular free-energy transducer, an immediate question concerns the kinds of useful energy which are transduced. It is quite clear, that mechanical (i.e., kinetic/-potential) forms are involved /22/. Considering, the predominance of activated protonic/electronic states during catalytic events at the active center, it is natural to ponder a role of the protein molecule as a protical/electrical transducer. There is an increasing indication, from both experimental and theoretical fronts, that such transduction does, indeed, occur in individual proteins, and that it may be involved in catalytic processes /22,23/. Interest in this type of energy transduction is heightened further, when we relate the cytological juxtaposition of organized enzyme systems and sites of protonic/electronic sources (see Section II). It is quite plausible that, in many cases, the enzyme molecule not only connects the active center mechanically to the surroundings, but, also, protically/-electrically.

Let us look at elements of protein structure which suggest such energy-transduction design. The obvious possibilities are regions of local secondary structure, viz., α-helix and ß-structure. First, we consider the α-helix. The α-helix possesses a marked dipole moment, originating in the alignment of the dipole moments of the individual peptide units /25/. The helix dipole generates a considerable electric field, whose strength increases with helix length - particularly in a medium (viz., the interior of a globular protein or a membraneous phase) with low dielectric constant.

Hol et al. /25/ suggested that the electric field of the α-helix is "a significant factor which must be included in the discussion of the properties of proteins." The previous authors proposed three possible functional roles of this field in enzyme activity: i) binding of charged substrates or coenzymes, ii) long-range attraction and/or orientation of charged substrates, and iii) contribution to catalytic events. Several, widely diverse types of enzymes are known, from structural studies, to bind negatively-charged compounds at the N-terminal region of an α-Helix. Moreover, Hol et al. /25/ specify a number of different kinds of enzymes (e.g., NAD-dehydrogenases), whose active sites are known to be located at the N-terminus of an α-helix. Moieties at such helix termini appear to function in proton-transfer networks which "activate" (deprotonate) catalytic nucleophiles and/or in stabilization of negatively-charged (e.g., oxyanion) transition states.

Other, interesting roles of α-helical elements have been suggested. Types of α-helices may constitute the framework through which proton transport occurs in the well-known, membraneous proton "pumps" /26-28/. Also, the α-helix seems appropriately designed for vectorial conduction of phonon-like energy impulses, from sites of energy release (e.g., ATP hydrolysis, redox reactions) to sites of utilization /29/.

Regular ß-structures in some enzymes (e.g., serine proteases) most likely contribute to local fields, involved in substrate-binding and/or catalytic events. It appears, that ß-structures provide an additional way of integrating the substrate into the protonic "circuitry" of the protein matrix /24/.

Another structural array, which may be particularly important in proton semiconduction (see Section IV), is a hydrogen-bonded network, formed by the interdigitated R-groups of parallel α-helix segments or parallel ß-structures /30/. Obviously, such a network could function efficiently as a proton conductor only within a relatively hydrophobic environment (viz., a membrane phase or inside a globular protein). As shown by Nagle et al. /31/, this design can be a modus operandi for proton "pumps", protochemical reactions, and, even, proton-driven mechanical work (e.g., cyclic conformational changes, generated by migrating "faults" which follow proton "hopping" - see Section IV).

Long-range, mobile protonic states are finding increasing relevance to many cellular processes, stemming from the original pioneering suggestion as to the role in electron-transfer phosphorylation /17/. As indicated in Section II above, this kind of energy continuum is emerging as a unifying theme in cell metabolism. The importance of this theme demands, that we begin looking at enzyme structure/function with new perspectives. Relating to basic enzyme action, as expressed by Wang /23/, such protonic states may represent "a broad catalytic principle which transcends the idiosyncracies of individual enzymes." The three enthalpic catalytic factors discussed above may all be subservient to this "principle". That is to say, facilitated proton transfer, in the enzyme-substrate complex, plays a key role in most enzyme catalyses. [This is obvious in the case of acid-base catalysis. It is more subtly involved in "rack" and covalent mechanisms. For, nucleophilic (e.g., hydroxyl, sulfhydryl) groups at the active center usually require "activation" by proton shuttles or charge-relay networks.].

A similar notion has been advanced forcefully by Metzler /24/, in a recent review on tautomerism in enzyme catalysis. As noted by the previous author, "the rapid equilibration of isomers in which one or more hydrogen atoms change positions is a prevalent phenomenon among biochemical substances." He proposed that enzyme molecules have evolved generally to conjugate with such protonic/electronic chemical configurations, and that tautomeric effects in enzyme-substrate complexes probably play basic roles in catalysis and regulation. As discussed by Metzler /24/, proteins contain many tautomeric groups, such as the peptide/amide groups (as discussed above) and many of the amino-acid side-chains, as well as bound coenzymes. The non-polar nature of the protein interior favors the formation of extended hydrogen-bonded networks (and ensuing charge mobility) among these various groups. Charge transfer through these chains is likely to be fundamentally important to enzyme action. In particular, addition or removal of a proton, at one end of a chain of such hydrogen-bonded groups, will produce a local field-effect at a distance (and vice-versa), via the "continuum" of the network /see also ref. 32/.

Superimposed on these designs may be long-range electronic semiconduction states which may be important in some enzymes, especially in organized states in vivo /33/. Various modes of electronic (and electron-phonon) coupling between enzyme and substrate have been offered as theoretical possibilities /34,35/.

B. Protein Dynamics

As indicated in Section III.A. above, there has been an increasing awareness in recent years, of the role of the protein matrix as a "mechanical mediator" between the enzyme active-center and the thermal properties of the bulk medium. Naturally, the question arises, as to the interaction between excited conformational (e.g., vibrational) states and excited protonic/electronic states in the enzyme molecule. This consideration is particularly important, when one notes that virtually all enzymatic mechanisms involve some kind of mechanical force, acting through electrostatic interactions, generated across chemical bonds in the substrate.

Recently, Welch et al. /22/ reviewed various theoretical models, which propose that spatiotemporal ordering of the fluctuational behavior of the protein molecule serves an integral role in enzyme catalysis. A notable principle, emerging from all of the models, is the unity of the enzyme molecule and the ambient medium. We find that biological evolution has apparently come to grips with the random field in the medium, by designing a macromolecular structure which is capable of "borrowing" and collimating that random energy-source in an anisotropic fashion.

Notwithstanding, there is something deceptive and misleading about this developing picture. Again, we find the taint of the reductionistic theme. Accepting the premise that most (if not all) enzymes of intermediary metabolism operate in organized regimes in vivo (see Section II), there is a certain degree of artificiality in our construction of general theoretical models which attempt to "unify" the dynamical protein molecule with a bulk aqueous medium. We contend, that by probing the above theoretical models more deeply, we discover a hint - from the fluctuational behavior of the protein in bulk solution - as to the mode of operation in the organized states.

External and internal hydrogen-bonding plays an integral role in the dynamic tertiary (and quaternary) structure of globular proteins /36,37/. Following tenets of certain of the dynamical enzyme models /22/, one sees that gaps ("faults", "defects") in the internal-bonding arrangement might elevate locally the free-energy of the system - electrostatically and mechanically, as well as "protonically" (see Section III.A). Catalytic functions would, then, depend on precise, internal "fault" states /36,37/. Such "faults" (or "defects") arise, in a protein dissolved in aqueous solution, in conjunction with binding-relaxation of bound water, fluctuating proton-transfer processes and charge-density fluctuations at the surface, etc. (Inside the protein, these "faults" can migrate, for example, by proton-hopping /30/. The energy required for a single "hop" is about kT, i.e., thermal energy.) Thus, isolated in bulk solution, the internal-defect pattern of a given protein is subject to a random generator from the solute/solvent system. The catalytic-turnover-numbers for isolated enzymes in vitro reflect this random field.

A major thesis in this article, is that mobile protonic states in organized enzyme systems in vivo are involved in such phenomena as protochemistry, substrate translocation ("channeling"), and coordinate "energization" of enzyme proteins. As noted in Section III.A, while the active-site configuration dictates the energetic character of the enzyme-substrate transition state, it is the protein matrix (in dynamic interaction with the surrounding medium) which determines the rate at which the system reaches that state. We propose, that proton-conformational-interaction (PCI - a term coined by Volkenstein /34/) constitutes a major mechanism by which enzyme-substrate complexes reach requisite transition states.

When "hooked-up" specifically to the "energy continuum" of a proton-motive-force (e.g., at a membrane interface), the action of a given enzyme may be "driven" dynamically by a steady-state proton flow (as well as by local field effects). This flow would serve, in some instances, a "stoichiometric" role, as a source/sink of protons for protochemical events at active-centers /23/ - with the protein matrix functioning as a conductor. And, generally, the "energy continuum" of the mobile protonic state, by virtue of proton-conformational-interaction (PCI), would serve to "energize" the enzyme-substrate complex (i.e., toward defined transition-states) - with the protein matrix functioning as a free-energy transducer.

Again we note the possible relevance of mobile electronic states, here in relation to protein dynamics. Interaction of electronic and nuclear degrees of freedom is well-known in solid-state physics. An example is electron-phonon coupling in crystal-lattice dynamics. The roles of such modes in enzyme action have been discussed by others /34,35,38/. And, as evident from the contributions in this volume from Prof. Fröhlich and others, excitation of coherent electrical (e.g. dipole) vibrations in proteins (via a phonon-condensation phenomenon) may play a role in conformational dynamics of enzyme action in structured regimes. (Indeed, there may exist in the living cell a duality of protonic/electronic "energy continua" which bear upon enzymatic processes /39/.)

IV. THE PROTON-SEMICONDUCTION MODALITY IN INTERMEDIARY METABOLISM

Let us now exceed the level of the individual enzyme molecule and explore holistically the manner by which the mobile protonic state ("proticity") can actually integrate the multienzyme systems of intermediary cell metabolism. First, we shall look at the "band structure" of the semiconduction mode.

A. Band Structure

Following the discussion in Section II above, we assume that most (if not all) of intermediary metabolism operates in a (quasi-) solid-phase environment. Whereby, the associated protein-protein or protein-particulate assemblages are amenable (approximately) to modeling as solid-state matrices. Many of the basic ideas on solid-state protonic semiconductors originated with studies on ice. This unique hydrogen bonded structure has been a model system, by virtue of the marked mobility of protons therein /30/. Nagle et al. /30,31/ have developed a model of proton semiconduction, in protein-membrane arrays, which is apropos of our design herein. We follow their inciteful approach in extenso in the ensuing treatment.

Nagle and Morowitz /18,30/ postulate as a new element in bioenergetics, that the "universal energy intermediate" in the living cell is a chain of hydrogen bonds with an extra proton and a potential barrier on each end. The injected protons can come from 1) chemical reactions, 2) a reservoir of protons (or hydronium ions) at high electrochemical potential, or 3) redox (or photo-redox) processes. Such a network might be formed via parallel α-helical or parallel β-structural elements, as discussed in Section III. Calculations /31/ show that such hydrogen-bonded chains can rapidly transport protons, from high potential to low potential, with little loss in energy. Such high-energy protons, before being released into solution with low electrochemical potential, could be made to do work. The actual free-energy change depends on such factors as the electrical field on the system, the difference in dielectric properties between the bulk and the conducting medium, and structural asymmetries in the hydrogen-bonded network. The energy difference could be as high as 0.6 kcal/bond-mol, e.g., 6.0 kcal/mol for a chain of 10 bonds /30/.

If this energized network is to perform useful work, there must be control over the presence of protons and faults in the system. This can be achieved by various gating mechanisms at the ends of the chains. Following Nagle and Morowitz /30/, one plausible mechanism involves a simple, cyclic alteration in the hydrogen-bond lengths accessible to a migrating proton or fault.

Structural/energetic asymmetry in the hydrogen-bonded chains can convert the system from a "passive" proton wire to an "active" device. Here, the passage of protons and faults along a chain is intimately coupled to conformational changes in the protein containing the chain /cf. ref. 28/. Such a device can function as a unidirectional proton pump or as a motor (producing cyclic conformational changes in the protein) (see Section III.B). Most importantly, from the biological viewpoint, the energy per cycle in these devices is tunable by the protein within wide ranges. And, in general, a given protein might contain more than one "active" hydrogen-bonded chain, which circumstance greatly increases the overall free-energy transduction.

Notably, the individual proton/fault - transfer steps only involve about kT in energy, and, thus, can be thermally (albeit randomly) activated. This explains why many enzymes, which might be localized in organized states in vivo, can work (at least, in some fashion) when isolated in bulk aqueous solution (see Section III.B).

B. Organized Enzyme States

Nagle and Morowitz /30,31/ applied this proton-semiconduction model to proton-pumping, as found in mitochondrial, thylakoid, and bacterial-membrane systems, as well as to the "protochemistry" of ATP synthesis. However, the previous authors alluded to more general roles. Of course, a major requirement in this design is a (relatively) hydrophobic medium for the "proton wiring", with possible participation of structured water. The bulk aqueous phase is too dissipative.

We would like to employ the tenets of the Nagle-Morowitz model, in the analysis of the functional modes of organized-enzyme states in the living cell. In so doing, we make the following postulates:

1) There exists, in the solid-phase regions of the cell interior, a mobile-protonic-state (continuum)/7/. This state is maintained by local redox (or photo-redox) processes or other proton-injector sources (e.g., H^+-ATPases). The conduction pathways for the proton flow may entail structured water and/or specialized network of proton-transfer proteins /40/. The protonic "continuum" may extend into the bulk phase, via the proteinaceous microtrabecular lattice (see Section II).

2) Many organized multienzyme systems in vivo are "hooked-up" functionally to particulate structures, so as to be "plugged" into the protonic continuum generated in the solid-phase regions.

3) Certain hydrogen-bond systems in the individual enzymes are designed to be wires, which function in PCI-energization of enzyme-substrate complexes (see Section III.B).

4) Enzymes, whose catalytic processes involve transient proton-transfer events in transition-state stabilization (e.g., acid-base catalysis), contain "active" wires which "energize" the catalytic-center "protodes" according to the ambient, mobile protonic-continuum (see Section III.A).

5) Enzymes, whose chemical reactions involve net proton consumption or production, contain "active" wires which interconnect the chemical reaction stoichiometrically to the ambient protonic-continuum (see Section III.A). Such enzymes represent local sources and sinks on the continuum (proton reservoir). Thus, protons may be injected or withdrawn by chemical reactions located at the termini of hydrogen-bonded chains (by analogy to ATP synthesis/hydrolysis in the H^+-ATPase /18/).

6) Inter-protein "wiring" (proton pathways) of the localized multi-enzyme assemblages is designed, in some cases, so that the mechanical translocation of intermediate substrates can be driven by proton-flow (i.e., PCI mechanical work - see Section IIIB). Hence, substrate-transfer ("diffusion") along the enzyme sequence can be coordinated with the conformationally-linked dynamics of the catalytic processes of the individual enzyme reactions by virtue of the universal mobile-protonic continuum to which the entire multienzyme assemblage (as a unit) is subservient.

7) Local electrical fields (at the particulate-cytosol interface) play a variety of functional roles. For example, they would aid in the proper alignment of many membrane-associated enzymes, in substrate-translocation, and in dictating the dynamics of proton flow.

Let us elaborate on the nature of these postulates. Postulate 1 is the conceptual basis for the entire picture, as discussed in Section I (see ref. 7). Proton-conducting pathways, along the planes of energy-transducing membranes, have been proposed by a number of workers (40). The idea, of a ("protoneural") network of proton-transfer proteins (see Dr. Kell, this volume), is particularly attractive. Importantly, such a scheme could explain how the system controls (gates) the flow of protons to particular "sinks" according to the energetic matching of the network conducting-proteins with specific enzyme proteins which utilize the flow. Insulation of the proton flow, along hydrogen-bonded chains, from protein to protein would be aided by the ambience of structured water at the surface of the particulates. Continuity would be ensured, though, by direct protein-protein interaction, involving quaternary contacts. As is well-known, such interactions usually involve hydrogen-bonding across an interface which is highly hydrophobic in nature /24,41/. Such interconnection may be important in allowing the mobile-protonic-state to extend into the bulk, via the microtrabecular lattice. A similar role of the lattice, in electron semiconduction, was suggested by Lewis /33/.

Postulate 2 implies, that enzymes have "social sites" /6/, which recognize binding sites on the particulates (as well as sites on other enzyme-proteins in the assemblage). In addition some internal amino-acid sequences (e.g., α-helical regions - see Section III) may specify the alignment of the enzymes with the local electrical-field structure /42/.

Postulate 3 follows from the discussion in Section III.B above. Here, we envisage the "energy continuum" of the mobile-protonic-state as coordinating the PCI-energization of the multienzyme assemblages. Also, postulate 4 comes directly from the analysis in Section III.A. These two postulates relate the role of the "energy continuum" in facilitating (and coordinating) the entrance of the individual enzyme-substrate complexes into their respective transition states.

Postulate 5 is based on the "proton-well" picture of H^+-ATPase-catalyzed ATP synthesis /18/. Nature has developed a very efficient and refined scheme, here, for coupling a chemical reaction stoichiometrically to an "energy continuum". The proton-motive-force is a ubiquitous entity in the living cell. Following Nagle et al. /18,30/, we propose, that such a coupling mechanism may be involved in many other proton consuming/producing enzyme reactions in the organized state in vivo /7/.

Concerning postulates 4 and 5, we see that proton abstraction or donation at the active site, depending on the chemical nature of the catalyzed reaction, would be facilitated by the relative orientation of the conducting (asymmetric) hydrogen-bonded chain(s) in the enzyme-protein with respect to the vector field of the local proton gradient and the vector of the local electrical field /31/. For reaction processes requiring proton-donation, the local residence time of the proton, emerging from the protogenic site in the enzyme active-center, is an important question. In this regard, Gutman /43/ has suggested, based on experimental findings, that such an enzyme microcavity "can serve as an efficient storage system, capable of accumulating more than one proton per catalytic event and utilizng their electrochemical potential as a driving force". Without such localized proton "channeling," the activity of a given enzyme molecule would be dictated by random, thermal fluctuations of the bulk pH - which, in the microenvironments _in vivo_, can be significant. (A similar consideration, in the opposite sense, holds for active-center groups which must be maintained _de-protonated_ for enzyme activity). (See Section V.)

Postulate 6 represents the metabolic counterpart of the well-known, proton-mediated membrane transport process. For such a scheme to be operative in the multienzyme systems in the solid-phase regions _in vivo_, there must be rather close juxtaposition of the various enzyme proteins. The issue here is one of "channeling" of intermediate substrates. Most of intermediary metabolism must function under conditions of efficient compartmentation /5/. The most expedient means of achieving this would involve a precise processing of _one molecule_ of the "pathway substrate" at a time along a sequentially-acting multienzyme complex. _In vitro_ studies with isolated multienzyme complexes have demonstrated the reality of this processivity ("channeling") /5/. And, such a compartmentation view is consistent with accumulating experimental measurements _in vivo_, indicating that substrate-enzyme molar ratios are often around _unity_ (and, in some cases, less than unity).

The alternative to this extreme, _holistic view_ of intermediary metabolism is the _reductionistic idea_, that the steady-state pool levels of intermediate substrates are at the whim of K/V ratios of the _individual_ (non-interacting) _enzymes_ /44/. Although one might suppose the existence of physical (e.g., viscosity) barriers on the out-diffusion of intermediate substrates in localized microenvironments /45/, it is difficult to imagine how such a metabolic process could operate without inefficiencies (e.g., leakage, side reactions).

Accepting the view that many of the multienzyme systems, organized in the solid-phase regions _in vivo_, actually form _protein-protein complexes_ (generating two-dimensional "crystalline arrays" on the particulate structures - see Section II), we must inquire into the energy source for the putative substrate-translocation phenomenon. Ultimately, this energy must appear as mechanical/conformational work, in the site-to-site translocation process. The energy can come from one (or a combination) of two possible sources. First, there is free energy derived from the _bound chemical subsystem_. Virtually all chemical reactions yield products, initially, with a nonequilibrium distribution of vibrational and, in some cases, electronic energy. The fate of such states in bulk solution is rapid (e.g., collisional) relaxation. However, within the confines of a multienzyme cluster, part of this energy may be retained within the protein superstructure and "used" in subsequent catalytic processes /5,6/. Substrate-translocation, among interacting protein molecules in a multienzyme assemblage, may be facilitated by this kind of transduction.

Secondly, the energy (motive force) driving the translocation might come from an external source. One possibility, proposed by Fröhlich /42,46/, involves the participation of Coulombic forces generated by excitation of dipole states in the organized enzyme assemblage (see Section III.B). Additionally, we suggest that, within protein-protein complexes in solid-phase regions in vivo, PCI-dynamics may play a significant role in facilitating the processing of intermediate substrates. This represents a possible corollary of postulates 3-5 above. These postulates imply, that the progress of the bound substrate along its prescribed chemical-reaction profile is coordinated with a local proton flow. This flow would act to drive the conformational changes occuring during enzyme catalysis in the manner of a cyclic engine /28,30,31/. (This idea is consistent with the two-state model of ATP-linked ion pumps, suggested by Hammes /47/.) The most expedient operation of a multienzyme array would entail "cyclical matching" of sequential enzymes, such that one enzyme in the aggregate is at the beginning of a "conformational cycle" at just the moment when the preceding enzyme-product complex is a the end of a "cycle". Optimal conditions for efficiency in energy flow, as well as in metabolite "channeling", would obtain, provided that substrate-translocation is coordinated with the proton flow in the enzyme ensemble. Interestingly, a recent study with aldehyde dehydrogenase /48/ revealed a stoichiometric proton release, from some ionizable group on the protein, triggered by a substrate-induced conformational change. (Details of this process have been discussed by Berry et al. /39/).

This translocation modality would demand "energetic matching " of the organized enzymes, according to the "protoneural hypothesis" of Kell /49/. In addition, it would require a degree of complementarity between a given enzyme and the enzyme-product complex of the preceding metabolic step (as opposed to the free product molecule). This notion was proposed by Friedrich /50/, as part of his "dynamic compartmentation" model. (Experimental support for this model has come, for example, from studies of glycolytic enzymes /51/.)

As evinced from studies on energy-transducing membranes /40/, proton translocation and local electric fields are interdependent. Regarding postulate 7, Berry et al. /7,39/ have considered a number of possible field-effects on the orientation of adsorbed enzymes, local proton flow, catalytic events, inter alia.

V. IMPLICATIONS

Multienzyme systems adsorbed to subcellular particulates may potentially be subject to a number of energy-transduction modes. Indeed, some of these modes have been treated in other contributions to the present volume. We have focused only on aspects related to the "mobile protonic state." Our aim has been to draw attention to a possible physiological connection, between the central bioenergetic concept of "proticity" and the universal role(s) of "mobile protons" in enzyme structure and function. It is quite plausible, that our mobile-protonic mode is inextricably associated with other transduction modes. Thus, it may be somewhat artificial, to dissect this as a distinct phenomenon. The possibility of such a unification remains to be explored.

Experimental evidence, in support of our holistic idea, is presently circumstantial and equivocal. Nonetheless, we believe the notion to be based on sound physicochemical and biological principles. And, the number and diversity of biochemical/biophysical concepts which are potentially consolidated seems beyond fortuity. Moreover, the traditional reductionistic manner of extracting and studying individual enzymes in vitro is biased against any long-range, energy-transduction modality. Removal of this bias may pave the way for empirical approaches which will recognize the synergistic nature of living systems - the whole is more than the sum of the parts. In the present context, it is clear that the course will require a combination of experimental methodologies from classical enzymology and contemporary bioenergetics.

Our "protochemical picture" bears many implications for biochemistry and cell biology. A number of important considerations have been stressed by Berry et al. /7,39/. These include consequences in such areas as metabolic regulation, hormonal action, and energetic efficiency of metabolism. In particular, the previous authors stressed the potentially erroneous way in which steady-state mass-action ratios are determined in vivo. Many metabolic (e.g., kinase, dehydrogenase, ATPase) reactions involve a proton as reactant or product. In free-energy calculations we customarily "ignore" the protons, assuming that they equilibrate with the bulk. Our "proto-chemical picture" suggests, that this approach gives false quantitative results for a host of cellular processes.

Here, we would like to note additional implications:

1) Local pH. The concept of "pH" relates to statistical, bulk-phase properties. As is obvious from Section III, the structure and function of most enzymes would be sensitive to pH changes. This consideration raises major questions, regarding operation of enzymes in organized states in vivo. In subcellular microenvironments the usual definition of "pH" breaks down, and fluctuations in proton concentration (activity) predominate. The attendant, serious implications for cellular enzymatic processes have been emphasized by a number of workers /14,52,53/. Clearly, this problem is obviated, if localized enzyme systems are "plugged into" a mobile-protonic-continuum, according to the postulates of Section IV.

2) Evolutionary considerations. We suggest, that the central roles of mobile protons in enzyme structure and function reflect the nature of chemical catalysis in the primordial conditions of prebiotic evolution. Dating from the early work of Bernal, it is widely held that prebiotic reactions took place on clay surfaces in aqueous media /54-56/. Clays are abundant, exist in great variety, and are capable of concentrating (adsorbing) small molecules from aqueous solution. Many clay minerals behave as effective acid catalysts, with unusual proton-donating properties. This is thought to result from water molecules which interact strongly with the clay surface. The bound (polarized) water apparently plays a crucial role in proton-transfer events during the local chemical reactions. Accepting the premise that biological evolution is an anagenetic process, we follow the suggestion of Good /56/, that present-day enzymes are "micro-colloids" reminiscent of primordial state. (Hence, the importance of hydration, observed for most enzymes in vitro.) Some (e.g., digestive) enzymes have evolved to function in bulk solution and are subject to thermal "energization" (e.g., by random proton-transfer processes) in the bulk medium. Whereas, many enzymes of intermediary metabolism may be designed to "plug

into" a coherent protonic state (perhaps of the form proposed above) maintained by the energy-yielding processes of the cell. In the latter case, one sees the necessity for individual enzymatic reactions to be subordinated into a coordinated unity.

Apparently, mobile protons (and conduction routes thereof) are intimately involved in the evolution and maintenance of the "living state". Enzymes, in their roles as highly specific and reactive "micro-colloids", appear to serve as intermediary agents in coupling mobile protons to chemical-reaction coordinates. As eobionts and early "life forms" became more complex, selection pressure may have dictated the creation of a means for homeostating the mobile-protonic-state, with the generation of a long-range energy continuum.

REFERENCES

1. Pagels, H.R. (1982). The Cosmic Code: Quantum Physics as the Language of Nature. New York: Simon and Schuster.
2. Zuidgeest, M. (1977). Acto Biotheor. 26, 30.
3. Lumry, R. and Biltonen, R. (1969). In Structure and Stability of Biological Macromolecules /S.N. Timasheff and G.D. Fasman, eds.), p.65. New York: Dekker.
4. Lumry, R. (1971). In Electron and Coupled Energy Transfer in Biological Systems (T. King and M. Klingenberg, eds.), p.1, New York: Bekker.
5. Welch, G.R. (1977). Prog. Biophys. Mol. Biol. 32, 103.
6. Welch, G.R. and Keleti, T. (1981). J. Theor. Biol. 93, 701.
7. Berry, M.N. (1981). FEBS Lett. 134, 133.
8. Zalokar, M. (1960). Exp. Cell Res. 19, 114.
9. Kempner, E.S. and Miller, J.H. (1968) Exp. Cell Res. 51, 150.
10. Coleman, R. (1973). Biochem. Biophys. Acta 300, 1.
11. Sitte, P. (1980). In Cell Compartmentation and Metabolic Channeling (L. Nover, F. Lynen, and K. Mothes, eds.), p.17. New York: Elsevier /North-Holland.
12. Srere, P. (1981). Trends Biochem. Sci. 6, 4.
13. Schliwa, M., van Blerkom, J., and Porter, K.R. (1981). Proc. Nat. Acad. Sci. USA 78, 4329.
14. Peters, R.A. (1930). Trans. Faraday Soc. 26, 797.
15. De Duve, C. (1964). J. Theor. Biol. 6, 33.
16. McClare, C.W.F. (1974). Ann. N.Y. Acad. Sci. 227, 74.
17. Mitchell, P. (1979). Eur. J. Biochem. 95, 1.
18. Morowitz, H.J. (1978). Amer. J. Physiol. 235, R99.
19. Gutfreund, H. (1976). FEBS Lett. 62, (Suppl.), E1.
20. Fersht, A. (1977). Enzyme Structure and Mechanism. San Francisco: Freeman.
21. Warshel, A. (1978). Proc. Nat. Acad. Sci. USA 75, 5250.
22. Welch, G.R., Somogyi, B., and Damjanovich, S. (1982). Prog. Biophys. Mol. Biol. 39, 109.
23. Wang, J.H. (1968). Science 161, 328.
24. Metzeler, D.E. (1979). Adv. Enzymol. 50, 1.

25. Hol, W.G.J., van Duijenen, P.T., and Berendsen, H.J.C. (1978). Nature (London), 273, 443.
26. van Duijnen, P.T. and Thole, B.T, (1981). Chem. Phys. Lett. 83, 129.
27. Krimm, S. and Dwivedi, A.M. (1982). Science 216, 407.
28. Dunker, A.K. (1982). J. Theor. Biol. 97, 95.
29. Scott, A.C. (1981). In Nonlinear Phenomena in Physics and Biology (R.H. Enns et al., eds.), p.7. New York: Plenum Press.
30. Nagle, J.F. and Morowitz, H.J. (1978). Proc. Nat. Acad. Sci. USA 75, 298.
31. Nagle, J.F., Mille, M., and Morowitz, H.J. (1980). J. Chem. Phys. 72, 3959.
32. Banacky, P. (1981). Biophys. Chem. 13, 39.
33. Lewis, T.J. (1979). In Submolecular Biology and Cancer (Ciba Foundation Symposium No. 67). New York: Excerpta Medica.
34. Volkenstein, M.K. (1981). J. Theor. Biol. 89, 45.
35. Conrad, M. (1979). J. Theor. Biol. 79, 137.
36. Lumry, R. and Rosenberg, A. (1975). Coloques Internationaux du C.N.R.S. (No. 246 - "L'Eau et Les Systèmes Biologiques"), p.53.
37. Ikegami, A. (1977). Biophys. Chem. 6, 117.
38. Caserta, G. and Cervigni T. (1974). Proc. Nat. Acad. Sci. USA 71, 4421.
39. Berry, N.M., Grivell, A.R. and Wallace, P.G. In Comprehensive Treatise on Electrochemistry, Vol. 10, Bioelectrochemistry (S. Srinivasan and Y.A. Bhizmadzhev, eds.). New York: Plenum Press, in press.
40. Kell, D.B. (1979). Biochim. Biophys. Acta 549, 45.
41. Hopfinger, A.J. (1977). Intermolecular Interactions and Biological Organization. New York: Wiley.
42. Fröhlich, H. (1975). Proc. Nat. Acad. Sci. USA 72, 4211.
43. Gutman, M and Nachliel, E. (1982). European Bioenergetics Conference (EBEC) Reports 2, 319.
44. Welch, G.R. (1977). J. Theor. Biol. 68, 267.
45. Welch, G.R. Somogyi, B., Matko, J. and Papp, S. J. Theor. Biol., in press.
46. Fröhlich, H. (1970). Nature (London) 228, 1093.
47. Hammes, G.G. (1982). Proc. Nat. Acad. Sci. USA 79, 6881.
48. Benett, A.F., Buckley, P.D., and Blackwell, L.F. (1982). Biochemistry 21, 4407.
49. Kell, D.B. and Morris, J.G. (1981). In Vectorial Reactions in Electron and Ion Transport in Mitochondria and Bacteria (F. Palmieri et al., eds.), p. 339. New York: Elsevier/North-Holland.
50. Friedrich, P. (1974). Acta Biochim. Biophys. Acad. Sci. Hung. 9, 159.
51. Weber, J.P. and Bernhard, S.A. (1982). Biochemistry 21, 4189.
52. McLaren, A.D. (1960). Enzymologia 21, 356.
53. Sols, A. and Marco. R. (1970)- Curr. Top. Cell. Regul. 2, 227.
54. Fripiat, J.J. and Cruz-Cumplido, M.I. (1974). Ann. Rev. Earth Plan Sci. 2, 239.
55. Mortland, M.M. and Raman, K.V. (1968). Clay and Clay Mineral 16, 393.
56. Good, W. (1973). J. Theor. Biol. 39, 249.

Indications of Optical Coherence in Biological Systems and Its Possible Significance

K. H. LI[1], F. A. POPP[2], W. NAGL[3], and H. KLIMA[4]

[1] Institute of Physics, Chinese Academy of Sciences, P. O. Box 603, Beijing, People's Republic of China
[2] Laboratory of Biophotons, P. O. Box 48, D-6521 Flörsheim
[3] Department of Biology, The University, P. O. Box 3049, D-6750 Kaiserslautern
[4] Atomic Institute, The Austrian Universities, Schüttelstraße 115, A-1020 Vienna

Introduction

The widespread, if not general phenomenon of "ultraweak" photon emission from living cells and organisms, which is different from bioluminescence[1)], has been extensively reviewed[2-4)] The measurements are carried out with a photomultiplier of high sensitivity in the range between 200 and 800 nm. The living material is kept within cuvettes in a dark chamber in front of the photomultiplier (for a more detailed description see refs.[3-5)]). With our equipment a photon current density of 1 photon/s/cm^2 can be detected at a significance level of 99.9 % within 10 hours. The uptake of count numbers within given time intervals and calculations are carried out with an interfaced computer.

The following characteristics of photon emission are now generally accepted:

(1) The intensity turns out to be of the order of a few up to some thousand photons per cm^2 per second.
(2) The spectral range spreads at least from infrared to ultraviolet.
(3) Proliferating cell cultures radiate more intensively than to those in which growth has ceased. The cells mainly emit in the G_1-phase.
(4) Dying cells exhibit a relatively intense photon emission, regardless of the cause of death such as refrigeration, heat, centrifugation or treatment with toxic agents.
(5) No agent is known which does not influence the photon emission.

However, nothing is known about the biological significance of this phenomenon. The interpretations reach from "imperfections" (= chaotic spontaneous chemiluminescence from various sources)[6)] and radical reactions (e.g. the dimol-singlet-triplet transition of oxygen $2O_2^S \rightarrow 2O_2^T$+h at 634.7 nm)[7-10)] to messages of genetic information[2, 11,12)]

With respect to Fröhlich's fundamental papers on energy storage and long-range coherence in biological systems[13,14)] and the basic work of Prigogine[15,16)] and Haken[17,18)], we feel that the aspect of coherence may play an important, if not the most essential

Coherent Excitations in Biological Systems
Ed. by H. Fröhlich and F. Kremer

role in order to describe and understand biological phenomena. This stimulated us to investigate some characteristics of biological photon emission with respect to its possible coherence. We will show that (1) there are some indications of at least partial coherence of this ultraweak radiation in biological systems, (2) we can describe these features in terms of coherent DNA-exciplex formation, and (3) a variety of biological phenomena can be understood within the framework of this model.

Some Physical Properties

The temperature-dependence of photon emission exhibits the same typical course as that of other physiological functions[3]. This indicates that the imperfection theory[6] cannot fit the situation. In contrast, we have to assume that the photon emission is either a product of a physiological process (e.g. the lipoxygenase reaction[7-10]), or a regulator of physiological functions. The activation energy due to thermal excitation corresponds to 0.4 to 0.6 eV.

As far as reliable results of the spectral distribution have been obtained, there is evidence that intensities of the same order of magnitude can be found within the range of at least 800 to 250 nm. This corresponds to occupation numbers f at about 10^{-20}, if we apply the photon emission to a stationary photon gas in phase space. The spectral distribution remains remarkably unaltered, even if the intensities increase up to a factor of 10^3 such as in the case of poisoning[3].

This observation is in accordance with some experiments, where an almost identical decay behaviour of all modes within the range of infrared to ultraviolet has been registered after exposure of cell populations to light illumination[3].

Of course, the photon intensity I(t) after excitation with light (which is switched off at a time t_0) drops down according to

$$I(t) = A(t+t_0)^{-\beta} \qquad (1)$$

where A and β are constants. $A(\beta)$ and β depend on the system under consideration. In most cases we find $1 < \beta < 3$.

The decay function (1) corresponds to an oscillator system with coherent (in this case frequency-stabilizing) feed-back scattering of radiation to its damped source. This can be seen by inserting the solution

$$x(t) = x_0(t)\cdot(1+d(t))\cdot\exp(i\omega_o t)$$

into the well known equation (2) of a damped classical oscillator.

$$\ddot{x} + 2\lambda(t)\cdot\dot{x} + \omega_o^2 x = 0 \qquad (2)$$

where the damping function $\lambda(t)$ accounts for the couplings within the system. Thereby, we confine ourselves to a definite class of slow-varying deformations d(t) of the source, namely to those expansions (or contractions) which are constantly put into the change of amplitude itself. This class is defined by condition (3)

$$\frac{\dot{d}(t)}{d(t)+1} = s \cdot \frac{\dot{x}_0(t)}{x_0(t)} \qquad (3)$$

where s and β are connected according to (3a), as can be seen by evaluation.

$$s = -\frac{\beta-3}{\beta-1} \qquad (3a)$$

A more detailed foundation, regarding quantum theoretical aspects, has been given in refs.[19,20].

Measurements of photo count statistics (PCS-experiments[21]) show that the probability $p(n,\Delta t)$ of registering n photons ($n = 0,1,2,...$) within a given time interval Δt is significantly different from a purely chaotic distribution, even for a multimode system with the highest possible degree of freedom[3,22]. On the other hand, the consistency of the results with a Poisson-distribution, which accounts for a coherent radiation field, cannot be refused.

Coherence of radiation within a medium requires some transparency[2]. If one violates, for example, a seedling at a certain position, at the same time a considerable increase of photon emission can be observed not only at the position of attack but also at the whole surface of the seedling[23]. This observation coincides with "light piping by plant tissues"[24], where a "remarkable degree of optical coherence" has been proved. The transport of coherent light takes place there without considerable loss not only at the membranes but through the whole cells.

Our investigations clearly show, in addition, that the DNA may represent the essential source (and sink) of "ultraweak" photon emission. This has been indicated by using ethidium bromide (EB) as a tracer molecule. EB intercalates into DNA without considerable reactions with other biomolecules. Intercalation of EB induces the unwinding of the DNA-superhelix-structures. After complete unwinding, a renewed winding with opposite rotational direction takes place at higher concentrations. We found that the intensity of photon emission from seedlings increased and decreased with the same dependence on EB concentration[25].

A Model

It is now generally accepted that exciplex formation takes place in DNA, which can compete with monomer emission and, in some cases, predominate even at room temperature[26,27]. Since exciplexes (excimers) exist only in the excited state, while the ground states are repulsive, such materials form a medium with negative absorption coefficient.

As an example, Figs. 1a and 1b display the energy diagram corresponding to a four-niveau system. k_4 represents the radiationless transition rate from level 4 to level 3, and k_2 is that from level 2 to 1. A, B are the Einstein's coefficients for spontaneous and induced transitions, respectively. n_i is the occupation number of energy level i, ρ represents the energy density of the radiation field due to transitions between level 3 and 2.

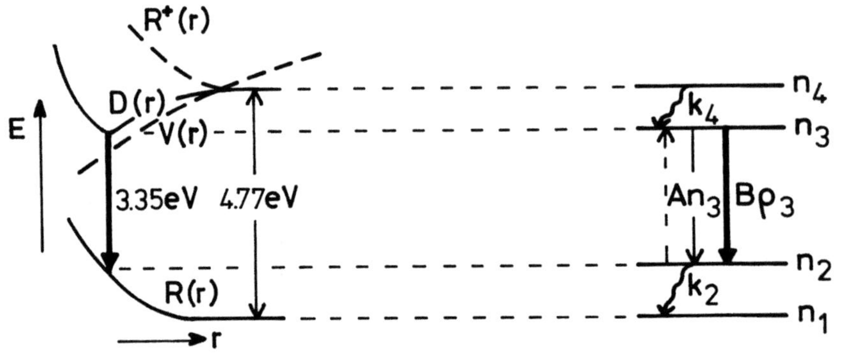

Fig. 1 a. Fig. 1 b.

We then obtain the following rate equations:

$$\dot{n}_4 = 1/2 \cdot Pn_1n_1^* - k_4n_4 \quad (4a)$$

$$\dot{n}_3 = k_4n_4 - (A + \rho B) \cdot n_3 + 1/2 \cdot \rho Bn_2 \quad (4b)$$

$$\dot{n}_2 = 2(A + \rho B)n_3 - k_2n_2 - \rho Bn_2 \quad (4c)$$

$$\dot{n}_1 = k_2n_2 - Pn_1n_1^* \quad (4d)$$

P is the "pumping" factor accounting for external energy supply due to collisions of n_1 unexcited and n_1^* excited monomers[28].

According to Haken's "order parameter method"[19] we get $\dot{n}_4 = \dot{n}_2 = 0$ for an "overdamped" motion due to k_4 and k_2, which accounts for rapid radiationless transitions within the same electronic potentials. Hence, we have

$$n_3 + 1/2n_2 = n = \text{const.} \quad (5a)$$

$$\dot{n}_3 - 1/2\dot{n}_2 = Pn_1n_1^* - 2(A + \rho B)n_3 + \rho Bn_2 \quad (5b)$$

After using the definitions

$$\sigma \equiv n_3 - 1/2\; n_2 \text{ and } \gamma(\sigma_o - \sigma) \equiv Pn_1n_1^* - 2An_3,$$

the equation (5b) can be rewritten in one of the coupled laser equations of O'Shea et al[29].

$$\dot{\sigma} = \gamma(\sigma_o - \sigma) - 2B\rho\sigma \quad (6a)$$

where σ_o is the equilibrium inversion of σ[18]. The other equation is obviously

$$\dot{\rho} = 2B\rho\sigma - K\rho \quad (6b)$$

with K representing the loss factor.

By use of the additional condition

$$n_1 + n_1{}^* + n_2 + n_3 + n_4 = N = \text{const.}, \qquad (7)$$

where N is the total number of bases, we obtain finally after expansion of ρ at about the initial value $\rho_\circ$:

$$n_3 = C_0 n + C_1 n_3 + C_2 n_3^2 \qquad (8)$$

with $C_0 \equiv B\rho_\circ$, $C_1 \equiv 1/2P(N-2n) - A - 2B\rho_\circ + B\rho' n$, $C_2 \equiv -2B\rho'$ and $\rho' \equiv (\frac{\delta\rho}{\delta n_3})\rho_\circ$.
We then have a metastable state of exciplex formation exhibiting laser action with a stabilized state at threshold[30).

The feed-back coupling of equ. (8) must not be overlooked in DNA. In particular, the appreciable occupation of level 2, which is different from the usual randomly distributed exciplexes in solution, plays a dominant role. It is also necessary to consider the interactions of phonons in the DNA-lattice system giving rise to transitions from, and to, level 2. As a consequence, the feed-back scattering of the radiation field may become coherent if, and only if, the photons are coupled with phonons of the DNA lattice. These interactions can be discussed in the framework of polaritons. This topic is now subject of our investigation and will be published elsewhere.

The experimental results and their interpretation apparently are consistent with coherent radiation of DNA. In addition, within the framework of our model, a variety of biological phenomena, such as cell differentiation and growth regulation, enzymatic activity, temperature dependence of physiological functions, active transport, pattern recognition (for instance: immunological response), chromatin condensation and cell adhesion, can be explained from a physical basis which escapes the difficulty of finding control mechanisms, which themselves do not need again a regulator ad infinitum [2,31)

References

1) Mamedov, T.G., G.A. Popov, V.V. Konev: Biophysics 14 (1969), 1102.
2) Popp, F.A. et al. (edts.): Electromagnetic Bio-Information, Urban & Schwarzenberg, München (1979).
3) Popp, F.A. et al.: Collective Phenomena 3 (1981), 187.
4) Ruth, B.: Dissertation, Marburg 1977.
5) Ruth, B. and F.A. Popp: Z.Naturforsch. 31c (1976), 741.
6) Zhuravlev, A.I. (ed.): Ultraweak luminescence in Biology, Moscow Soc.Naturalists 39 (1976), Moscow, P.17.
7) Boveris, A. et al.: Proc.Natl.Acad.Sci. USA 77 (1980), 347.
8) Lloyd D. et al.: Biochem.J. 184 (1979), 149.
9) Cadenas, E. et al.: Eur.J.Biochem. 119 (1981), 531.
10) Torinuki, W. and T. Miura: Tohoku J. exp. Med. 135 (1981), 387.
11) Popp, F.A.: Arch.Geschwulstforsch. 44 (1974), 295.
12) Kaznachejev, V.P. and L.P. Michailowa: Ultraweak radiation from cells as transmitter of intercellular information, Nauka, 1981.
13) Fröhlich, H.: Int.J.Quantum Chem. 2 (1968), 641.
14) Fröhlich, H.: Advances in Electronics and Electron Physics 53 (1980), 85.
15) Prigogine I. et al.: Physics Today 11 (1972), 23.

16) Nicolis, G. and I. Prigogine: Self-Organization in Nonequilibrium Systems, Wiley, New York 1977.
17) Haken, H.: Z.Phys. 181 (1964), 96.
18) Haken, H.: Synergetics, Springer, Berlin 1977.
19) Popp, F.A., K.H. Li: submitted for publ.
20) K.H. Li and F.A. Popp: Physics Letters A, in press.
21) Arecchi, F.T.: In: Quantum Optics (R.J. Glauber edt.), Acad.Press, N.Y. 1969.
22) Popp, F.A. et al.: In: Wiederherstellung und Erneuerung als Prinzipien der Organo- und Immuntherapie (K. Theurer et al. edts.), Enke, Stuttgart 1981.
23) Popp, F.A.: Unpublished Results.
24) Smith, H.: Nature 298 (1982), 423.
25) Rattemeyer M., et al.: Naturwissenschaften 11 (1981), 572.
26) Vigny, P. and M. Duquesne: In: Excited States of Biological Molecules (J.Birks edt.), Wiley, London 1976.
27) Morgan, J.P. and M. Daniels: Photochem.Photobiol. 31 (1980), 207.
28) Birks, J.P.: Rep. Prog. Phys. 38 (1975), 903.
29) O'Shea, D.C. et al.: Introduction to Lasers and their Applications, Addison-Wesley, London 1978.
30) Li, K.H.: Laser + Elektrooptik 13 (1981), 32.
Popp, F.A.: ibid., p.34.
31) Nagl, W. and Popp, F.A.: submitted for publication.

Self-focusing and Ponderomotive Forces of Coherent Electric Waves: A Mechanism for Cytoskeleton Formation and Dynamics

E. DEL GIUDICE[1], S. DOGLIA[2], and M. MILANI[1]

[1] Istituto Nazionale di Fisica Nucleare,
[2] Gruppo Nazionale di Struttura della Materia del C. N. R., Istituto di Fisica dell'Università, Via Celoria 16, I-20133 Milano

Coherent electric waves have been predicted to exist in biological systems.(1) We try to link the consequences of their propagation in the cell medium to some peculiar features of cell architecture and organization such as cytoskeleton formation and dynamics.(2) This will give a consistency argument in favour of the existence of coherent electric waves in living systems.(3)

The general ponderomotive force produced by an electric field $\vec{E}$ and a magnetic field $\vec{H}$ acting on a dielectric with polarization $\vec{P}$ and magnetization $\vec{M}$ is given by(4)

$$\vec{f} = \rho\vec{E} + \frac{1}{c}\vec{j} \times \vec{H} + (\vec{P}\cdot\nabla)\vec{E} + (\vec{M}\cdot\nabla)\vec{H} + \frac{1}{c}\frac{\partial\vec{P}}{\partial t} \times \vec{H} - \frac{1}{c}\frac{\partial\vec{M}}{\partial t} \times \vec{E} \qquad (1)$$

In a first approach we will drop the terms depending on the magnetization $\vec{M}$ (which is expected to be important only in a specialized class of cells) and we will also omit the linear term arising from the "true" current source $\vec{j}$. Therefore

$$\vec{f} = \rho\vec{E} + (\vec{P}\cdot\nabla)\,\vec{E} + \frac{1}{c}\,\frac{\partial\vec{P}}{\partial t} \times \vec{H} \qquad (1')$$

In eq. (1') we have the usual linear term $\rho\vec{E}$, the nonlinear term, called the "gradient force" $(\vec{P}\cdot\nabla)\,\vec{E}$, and the term of "radiation pressure" $\frac{1}{c}\frac{\partial\vec{P}}{\partial t} \times \vec{H}$. Since coherent waves can bring very large values of $\vec{E}$ (of the order of 10^7 V/m), the nonlinear terms of eq. (1') are expected to dominate on the linear one. Consequently, we will drop the first term in the right hand side of eq. (1'). The gradient term can be written as(5)

Coherent Excitations in Biological Systems
Ed. by H. Fröhlich and F. Kremer

$$(\vec{P}\cdot\nabla)\vec{E} = - \text{grad } p_o(\rho,T) + \frac{1}{8\pi} \text{grad } E^2\rho(\frac{\partial\varepsilon}{\partial\rho})_T + \qquad (2)$$
$$- \frac{1}{8\pi} E^2 \text{ grad } \varepsilon$$

where ε is the dielectric permittivity of the medium and $p_o(\rho,T)$ is the pressure in the medium in the absence of an external field, at the same volume and temperature T. The pressure gradient term, although relevant in many applications, can be disregarded in a microscopic approach. Moreover, we will use an approximation valid for gaseous media and dilute solutions (as the cytoplasm can be considered in a rough approximation):

$$\rho \frac{\partial\varepsilon}{\partial\rho} \sim (\varepsilon - 1) \qquad (3)$$

After trivial manipulations one gets from eq. (1')

$$\vec{f} = \frac{1}{8\pi}(\varepsilon-1)\nabla E^2 + \frac{1}{c} \frac{\partial\vec{P}}{\partial t} \times \vec{H} \qquad (4)$$

The first term in the right hand side of eq. (4) shows that a high power, strong directional electric wave

$$\vec{E}(\vec{r},t) = \frac{1}{2} \vec{A}(\vec{r},t) \exp\left[i(\omega t - \vec{k}\cdot\vec{r})\right] + \text{c.c.}$$

can produce condensation or rarefaction in a medium according to the sign of the coefficient of ∇E^2. This coefficient is a function both of the frequency ω of the beam and of the typical resonance frequencies ω_{ok} of the dielectric. It has been shown[6] that

$$\frac{1}{8\pi} (\varepsilon-1)\nabla E^2 = C \sum_k \left[\frac{\omega_{ok}^2 - \omega^2}{(\omega_{ok}^2 - \omega^2)^2 + \Gamma_k^2}\right] \nabla E^2 \qquad (5)$$

where C is some positive constant and Γ_k is the damping of the k-th oscillator. This gradient force is important when the incoming frequency ω approaches ω_o, one of the frequencies of the medium. An attractive (repulsive) force in the direction transverse to the propagation axis of the beam will then be present when $\omega_o > \omega$ ($\omega_o < \omega$).

The gradient force produces then a waveguide in the medium along the intense and directional electric beam, where a condensation of matter takes place. This process eliminates the geometric and diffraction divergences of the beam, leading to a very efficient propagation within the waveguide ("self-focusing").

In order to discuss this kind of propagation, the gradient force can be expressed in terms of the properties of the refractive index of the nonlinear medium[7]. On the basis of the Kerr properties of protein solutions[8], we will consider a "cubic" medium, i.e. $n = n_o + n_2|A|^2$ $(\varepsilon = \varepsilon_o + \varepsilon_2|A|^2)$.

From the Clausius-Mossotti equation for an isotropic medium

$$\frac{n^2-1}{n^2+2} = \frac{4}{3}\pi\rho\alpha \tag{6}$$

where α is the polarizability, we can express the variation Δn (at the lowest order) induced by an electric field $\vec{E}$ as

$$\begin{aligned} \Delta n &= (\Delta n)_\rho + (\Delta n)_\alpha \\ (\Delta n)_\rho &= \frac{1}{2} K_\rho \lambda |A|^2 \\ (\Delta n)_\alpha &= \frac{1}{2} K_\alpha \lambda |A|^2 \\ \lambda &= \frac{2\pi}{n}\frac{c}{\omega} \end{aligned} \tag{7}$$

K_ρ, K_α are the electrostrictive and ac Kerr constants respectively. $(\Delta n)_\rho$ is the variation of the refractive index due to electrostriction; $(\Delta n)_\alpha$ is the variation of the refractive index arising from ac Kerr contribution. It should be noticed(9) that only in the case of non-polar molecules it is possible to approximate the ac Kerr constant K_α with the contribution coming from the induced dipole moments $(K_1)_{dc}$ to the dc Kerr constant; this contribution is always positive whereas the dc Kerr constant K_{dc} can also be negative. For polar molecules with large dipole moments, K_α is in general smaller than the absolute value of the dc Kerr constant. Again we find as it appeared from eq. 2, that the propagation of an electric wave in a nonlinear dielectric is governed by an electrostriction term and by a molecular orientational one (Kerr term). From eqs. (2) and (7) the Kerr contribution appears to depend on ω and can be strongly enhanced when ω approaches a resonant frequency of the medium.

As far as the propagation of the beam is concerned, it can be shown(7) that eq. (7) implies the "self-focusing" of the beam, i.e. the beam shrinks and, after travelling in the medium a critical length

$$R_{nl} = \frac{a}{2}\left(\frac{n_o}{n_2|A|^2}\right)^{\frac{1}{2}}$$

(a being the radius of the cross-section of the beam), gets confined in a thin waveguide ("self-trapping") provided that its power be greater than a critical value

$$P_{cr} = (1.22)^2\lambda^2 c/256\, n_2$$

The above discussion compels us to conclude that coherent electric waves form in the medium a network of waveguides which in turn may attract "selectively" molecules in solution which have appropriate vibrational frequencies. Self-focusing strengthens very much the role of the gradient force by shrinking the beam and steepening its profile. Therefore molecules will experience stronger forces when the conditions for the beam self-trapping are met. An increase of concentration and an alignment of the solute molecules are then generated around the beam, resulting in a dynamic self-association, which is a requirement for polymerization. It has been recently reported[10] that polymerization can be actually induced by ordering the monomers.

At this point, let us turn to the radiation pressure term in eq.(4). The direction in which this force is exerted depends upon the longitudinal or transversal character of the time dependent electric field : a longitudinal wave produces a force transversal to the direction of propagation whereas in the transversal case the force is directed along the beam axis. As a consequence, in the first case the gradient force will change its magnitude; in the second case, molecules will also be pushed along the beam axis. The interplay between gradient forces and radiation pressure gives rise to an "optical mechanics"as shown by the experiments on laser induced"levitation" performed by Ashkin[11] and on laser isotope separation[12]: biological systems seem to be an excellent lab for such an"optical mecanics".The radiation pressure squeezes then the molecules, selected and attracted by the gradient force, in the waveguide and transports them along it.This squeezing enhances the probability that monomers ordered by the gradient force could polymerize. The contribution of the radiation pressure term to the assembly of the monomers could account for the dynamics of polymerization: the polymer can lose monomers at the end of the self-trapped beam while acquiring them at the opposite end.

The cytoskeleton consists of three different networks of filaments built up with protein (mainly actin and tubulin) polymers. These filaments exhibit a labile character, and their formation and dynamics strictly depend on a number of factors such as cell shape,timing, environment [13,14]. Moreover, a "treadmilling" behaviour has been observed[15]: monomers are assembled at one end of the filament and disassembled at the opposite end, giving rise to a unidirectional flow of matter. These features, insofar unexplained, can thus be accounted for by the ponderomotive forces generated by coherent electric waves propagating in form of filaments in the nonlinear medium.

An evidence for the general character of the proposed model is provided by the experiments on blood clotting[16]; filaments - the so-called contractils - appear when erythrocytes interact among themselves "à la Fröhlich" and an appropriate macromolecule is present in the suspension.

REFERENCES

1) H.FRÖHLICH, Advances in Electronics and Electron Physics (edited by L. Marton and C.Marton) vol. 53, p. 85 (1980)

2) I.I.WOLOSEWICK and K.R.PORTER, J.Cell.Biol. 82. 114(1979)

3) E.DEL GIUDICE, S. DOGLIA and M.MILANI, Phys.Lett. 85 A, 402 (1981); Physica Scripta 26, 232 (1982); Phys.Lett. 90 A, 104 (1982);in "The Application of Laser

Light Scattering to the Study of Biological Motion" (edited by J.C. Earnshaw and M.W.Steer(Plenum, London and New York (1982).

4) A.EINSTEIN and J.LAUB, Ann. Phys.(Leipz.) 26, 541 (1908)

5) P.MAZUR and S.R. De GROOT, Physica 22, 657 (1956)

6) G.A. ASKAR'YAN, Sov.Phys. JETP 15, 1088 (1962)

7) S.A. AKHMANOV, A.P. SUKHORUKOV, and R.V. KHOKHLOV, Sov. Phys. Usp. 93, 609(1967)

8) E.FREDERICQ and C.HOUSSIER, Electric Dichroism and Electric Birefringence (Clarendon,Oxford, 1973)

9) Y.R.SHEN, Phys.Lett. 20, 378 (1966)

10) T.FOLDA, L.GROS and H.RINGSDORF, Makromol. Chem., Rapid Commun. 3, 167 (1982)

11) A.ASHKIN, Phys.Rev.Lett. 24, 156 (1970)

12) N.V.KARLOV and A.M. PROKHOROV, Sov. Phys. Usp. 19, 285 (1976); N.G.BASOV, E.M. BELENOV, V.A. ISAKOV, E.P.MARKIN, A.N.ORAEVSKII, and V.I.ROMANENKO, Sov. Phys. Usp. 20, 209 (1977)

13) L. CARLSSON, L.E. NYSTRÖM, I.SUNDKVIST, F. MARKEY, and U. LINDBERG, J.Mol. Biol. 115, 465 (1977).

14) J.S. CLEGG, Collective Phenomena Vol. 3, p. 289 (1981)

15) R.L. MARGOLIS and L. WILSON, Nature 293, 405 (1981)

16) L.S. SEWCHAND, D. ROBERTS and S. ROWLANDS, Cell Biophysics, to appear; S.ROWLANDS, C.P. EISENBERG and L.S. SEWCHAND, J. Biol.Physics, to appear.

Specific Effects in Externally Driven Self-sustained Oscillating Biophysical Model Systems

F. KAISER
Institute of Theoretical Physics, University of Stuttgart, D-7000 Stuttgart

Introduction

Some years ago it had been proposed that the excitation of self-sustained oscillations (limit cycles) should be important for certain biological processes (1). Both, an increasing number of frequency specific effects in irradiated biological systems and some model calculations seem to support the limit cycle concept (2-5 and the contributions of Fröhlich, Grundler and Drissler in this volume). The present article is a brief review of simple models and deals with new aspects of the external drive of limit cycles.

The Models

Some results of the following models will be given.

Model I: *Externally driven generalized Van der Pol oscillator (6)*

$$\ddot{x} + \mu(-1 + x^2 - \alpha x^4 + \beta x^6)\,\dot{x} + x = F(t) \qquad \text{M I}$$

Without external drive, this equation exhibits two stable limit cycle oscillations.

Model II: *Coherent oscillation model (1, 6)*

$$\dot{\nu} = \gamma\sigma + \alpha A\sigma\nu + (c^2 e^{-\Gamma^2\nu^2} - d^2)\nu + F(t) \qquad \text{M II}$$
$$\dot{\sigma} = -\beta\nu - \alpha A\sigma\nu$$

With $F(t) = 0$, a limit cycle exists within a certain parameter region. The limit cycle contains both, a periodic chemical reaction and an electric vibration.

Concept and Consequences

Phase A: *Onset of a self-sustained oscillation ($F(t) = 0$)*

Necessary prerequisites for the creation of a self-sustained oscillation

Coherent Excitations in Biological Systems
Ed. by H. Fröhlich and F. Kremer

are nonlinear interactions within the system and dissipative processes. Under these conditions a periodic process can be produced at the expense of a non-periodic source of energy.

Phase B: *External perturbation of a self-sustained oscillation* *($F(t) = F_o \cos \lambda t$)*

step 1 | $LC(\omega_o) \rightarrow QP,\ NP \rightarrow LC(\lambda)$ |

The free internal oscillation (frequency ω_o) is perturbed by the external drive (strength F_o, frequency λ). With increasing F_o, the internal oscillation runs through a series of quasi-periodic (QP) and irregular (NP) states until it completely gets entrained (i.e. it oscillates with λ and a fixed amplitude, given by F_o). A sharp resonance peak results, provided a certain threshold for F_o has been reached. Specific results are given in Fig. (1-2), showing a strong dependence on λ and F_o.

step 2 | $LC(\lambda) \xrightarrow[F_o,\lambda]{} C$ |

The entrained limit cycle receives further perturbations and it can be driven into an unstable state (collapse). However the critical field strength $F_c(\lambda)$ leading to this collapse can be shifted to much higher values $\tilde{F}_c(\lambda)$, if in addition a static field F_1 is superimposed to the periodic field (F_1 has the tendency to create a stable nonoscillating state). The following bifurcation schemes result ($F(t) = F_o \cos\lambda t + F_1$; $F_1 = 0$ in the first sequence):

$LC(\omega_o) \rightarrow LC(\lambda) \rightarrow LC(\lambda,\lambda/2) \rightarrow NP \xrightarrow[F_c(\lambda)]{} C.$	S I
$LC(\omega_o) \rightarrow LC(\lambda) \rightarrow LC(\lambda/2) \rightarrow LC(\lambda/4) \cdots\rightarrow LC(\frac{\lambda}{2^n}) \rightarrow$ NP1 $\rightarrow LC(\lambda/6) \rightarrow LC(\lambda/12) \cdots\rightarrow$ NP2 $\xrightarrow[\tilde{F}_c(\lambda)]{} C.$	S II

(NP=nonperiodic state, i.e. irregular or chaotic; C=collapse). With the competition of F_o and F_1 the limit cycle runs through 2 series of period doubling states (P2, P4,... P128,... and P6, P12... P48...) and two nonperiodic situations before it collapses. No other stable state exists.

step 3 | C $\rightarrow$ pulse propagation |

The collapse of the oscillating system leads to the onset of specific types of travelling waves and signal propagation. The critical fields $F_c(\lambda)$ and $\tilde{F}_c(\lambda)$ are strongly dependent on the external frequency λ. It turns out that the route from the free internal oscillation (state of self-organization) to the collapsing state can be very different, depending on both, the specific limit cycle and the type of external perturbation.

For model M I one finds a transition from the free to the entrained oscillation and finally to the large amplitude oscillation (vid. figs.

1 and 2a). Under certain conditions the opposite behaviour occurs (fig. 2b). In summary, the following situation is given for M I:

step 1': $LC(\omega_0) \rightarrow QP,\ NP \rightarrow LC(\lambda) \rightarrow LC(\tilde{\omega}_0) \rightarrow QP,\ NP \rightarrow LC(\lambda)$

i.e. a superposition of 2 times step 1.

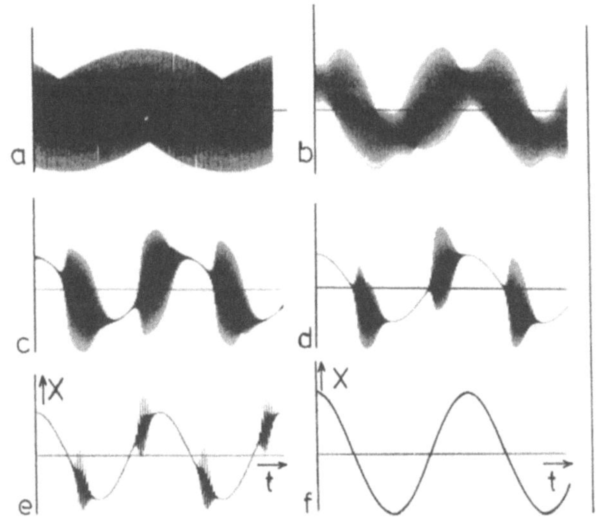

Figure 1. Competition between slow external and fast internal oscillation
Oscillation diagram (x as a function of time t) for MI with $\lambda \ll \omega_0$. F_0 increases from a to f, the internal oscillation gets more and more quenched, the system is completely entrained in f

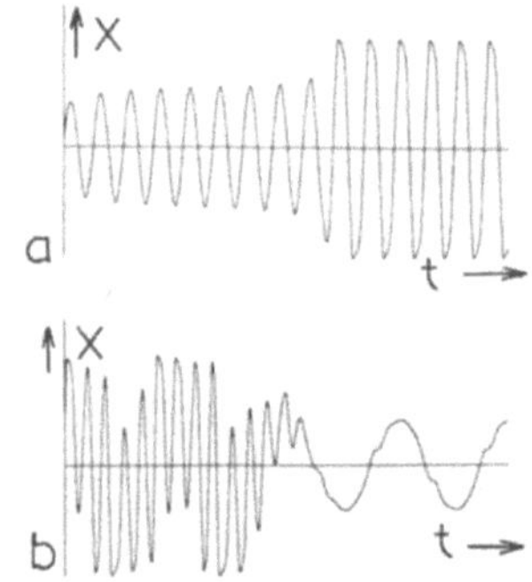

Figure 2. Oscillation diagram (MI)
a. $\lambda \approx \omega_0$, jump to large amplitude oscillation after a certain time of external drive
b. $\lambda \ll \omega_0$, jump to small amplitude oscillation with frequency λ

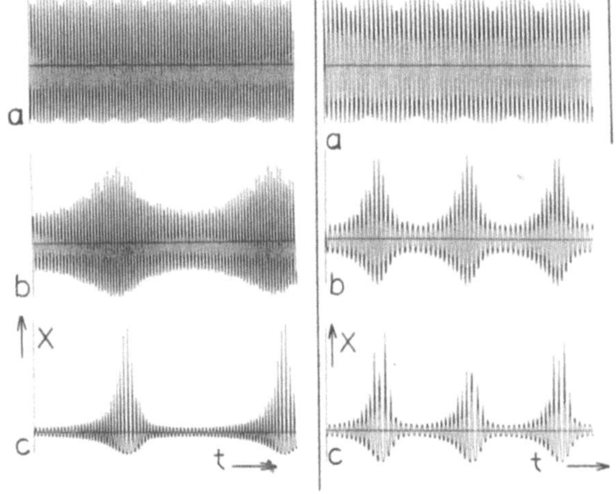

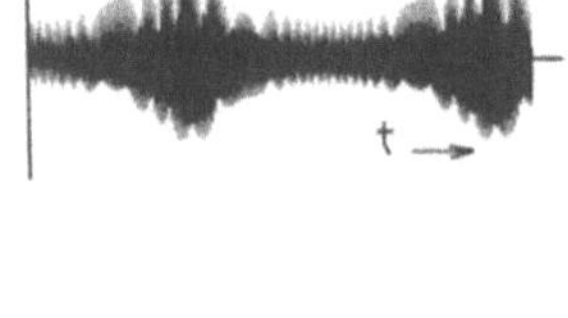

Figure 3. Collapse through beat oscillations
Oscillation diagrams (x as a function of time t) for MII. The external field strength F increases from a to c, thereafter the collapse occurs.
l.h.s.pictures : $\lambda \approx 1/10\ \omega_0$
center pictures : $\lambda \approx 1/4\ \omega_0$
r.h.s.picture : $\lambda \approx 1/20\ \omega_0$

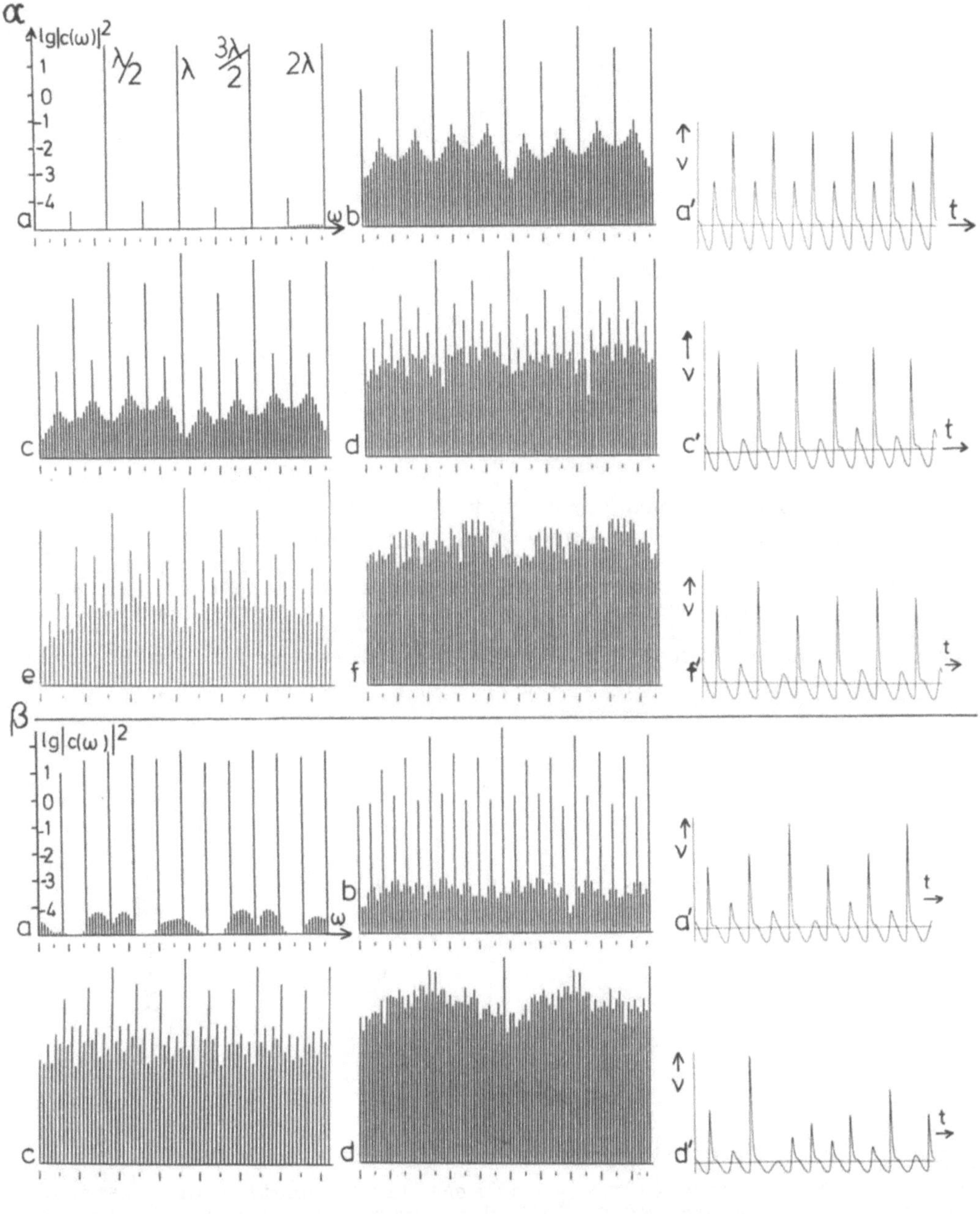

<u>Figure 4.</u> Bifurcation sequences

Power spectra ($|\lg c(\omega)|^2 - \omega$) diagrams for M II . $F(t) = F_o \cos\lambda t + F_1$. F_o increases from a to f (part α) and furthermore from a to d (part β). a', c', f' and a', d' are the corresponding oscillations ($\nu \equiv x$) to the power spectra a,c,f and a,d.

α : P2 P4 P8 P16 P32 NP1 (Pn = periodicity n)

β : P6 P12 P18 NP2 (NP = aperiodic)

(frequency range : 0 - 2λ ; 0 - λ in e (part α))

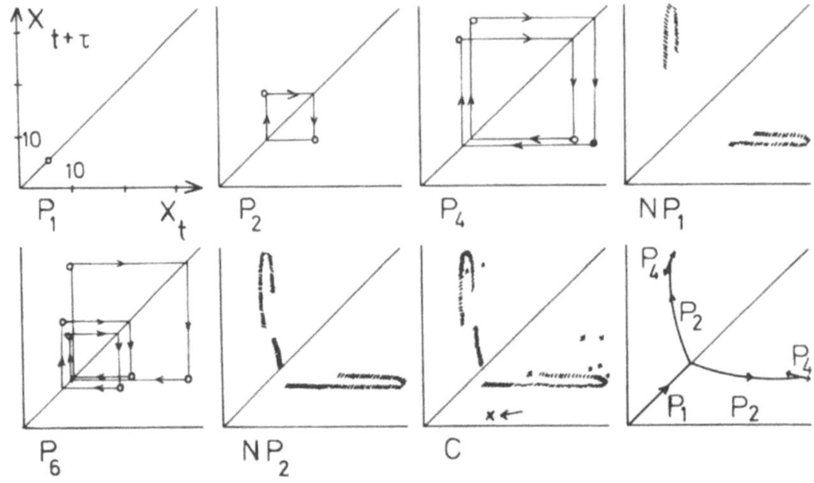

Figure 5. Stroboscopic phase portrait

$x_{t+\tau}$ as a function of x_t ($\tau = 2\pi/\lambda$) for different values of F_o. The pictures P1 , P2 , P4 , NP1 , P6 and NP2 belong to the corresponding pictures of figure 4 . C is the collapsing state, stars denote deviations from the stable amplitude regions, the star with an arrow shows the last amplitude before the collapse occurs. The last diagram of the series demonstrates the amplitude development for increasing F_o from P1 to P4

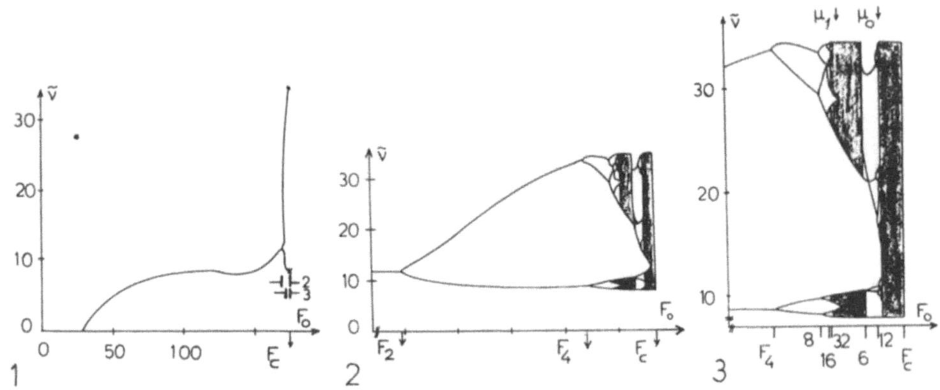

Figure 6. Bifurcation diagrams

Stroboscopic amplitude $\tilde{V}$ ($\equiv x_{t+n\tau}$, n large) as a function of the field strength F_o. The second and the third diagram are enlarged regions of the l.h.s. diagram , the F_o region chosen is marked with the numbers 2 and 3. F_1 and λ are fixed, the parameters are the same as in figs. 4 and 5. The numbers on the abscissa of picture 2 and 3 refer to the periodicities of figs. 4 and 5 , shaded regions are nonperiodic situations.

In figs. 4-6 the stroboscopic amplitudes are those values of the oscillations, for which the external periodic field has its maximum

The results for model M II are much more complex. An excitation with a very slow external frequency leads to a collapse of the oscillation via nonlinear beat oscillations (most of the amplitudes decrease, only a few grow up very strongly; vid. fig. 3). For $F(t) = F_o \cos \lambda t$ the sequence S I occurs, whereas for $F(t) = F_o \cos \lambda t + F_1$ the sequence S II is exhibited (vid. fig. 4-6).

The consequences of the different types of behaviour are:

- step 1 leads to sharp resonances (entrainment region) including threshold and time of irradiation effects,
- step 1' accounts, in addition, for frequency and intensity windows,
- S II exhibits both, an enlargement of the stability region and a rather complex behaviour with respect to the frequency of the resulting oscillations (period doubling...).
- step 3 (including the forgoing steps) may give an explanation of the frequency dependency and the extreme sensitivity of resulting effects.

Finally, the complicated behaviour of M I and M II and its strong dependence on F_o, F_1 and λ may serve as a basis for speculations on the occurence of extraordinary experimental results in irradiated systems.

References:

(1) Fröhlich, H., Neurosci. Res. Progr. Bull. 15, 67-72 (1977)

(2) Fröhlich, H., Advance in Electronics and Electron Physics, vol. 53, 85-151 (1980)

(3) Adey, W.R., Physiol. Rev. 61, 435-514 (1981)

(4) Kaiser, F., Z. Naturforsch. 33a, 294-304 and 418-431 (1978)

(5) Kaiser, F., ACS Symp. Series 157, 219-241 (1981)

(6) Kaiser, F., Theory of resonant effects of rf and mw energy in:"Biological effects and dosimetry of nonionizing radiation", p 251-282, Plenum Press New York (1982)

Addendum

A recent theory of Feigenbaum predicts that 1D unimodal maps should behave in a universal manner. Their dynamics run through a sequence of period-doubling bifurcations and chaotic states. In addition, the bifurcation parameter should asymptotically satisfy a recurrence relation, leading to a universal convergence rate (Feigenbaum's constant δ). Many examples agree with these predictions. It is remarkable that the externally driven self-sustained oscillator (M II) exhibits the same dynamics. The period-doubling sequence of S II is in good agreement with Feigenbaum's results, the bifurcation parameter F_o seems to converge to δ or at least to its vicinity.

Forces on Suspended Particles in the Electromagnetic Field

FRIEDRICH A. SAUER
Max-Planck-Institut für Biophysik, D-6000 Frankfurt 70

1. Introduction

There occurs the phenomenon of pearl chain formation above a critical field strength if a suspension of small particles (diameter of approximately 10^{-4}) is exposed to an electromagnetic field. For special cases when the medium and the particles show no dielectric losses, qualitative and semi-quantitative analysis has been given by Krasny-Ergen (1) and Saito (2).This analysis is based on relations for the body forces which are derived from energy principles in electrostatics and cannot be applied when dielectric losses occur (see for example Landau-Lifshitz (3)).

This paper reports the results of calculations for the force of interaction between two particles, when the medium and the particles have dielectric losses. These results enable us to calculate the trajectories of the particles in the process of pearl chain formation.

2. Calculation of Body Forces from the Electromagnetic Momentum Balance

Two balance equations can be derived with the help of the time dependent Maxwell equations; one for the electromagnetic field energy density $u_e = \frac{1}{2}\left(\vec{E}^T\cdot\vec{D} + \vec{H}^T\cdot\vec{B}\right)$ and the other for the electromagnetic momentum density $\vec{p}_e = \frac{1}{c^2}\left(\vec{E}\times\vec{H}\right)$.

In the following we make use of the nabla notation and $\vec{E}^T$ for example means the transposed vector. The symbols used are explained in the list of the symbols. We follow here the derivations given by Meixner (4) and De Groot and Mazur (5). With the space-time dependent Maxwell equations

$$\nabla\times\vec{H} = \vec{j} + \frac{\partial}{\partial t}\vec{D} \quad ; \quad \nabla\times\vec{E} = -\frac{\partial}{\partial t}\vec{B}$$
$$\nabla^T\cdot\vec{D} = \rho \quad ; \quad \nabla^T\cdot\vec{B} = 0 \tag{2.1}$$

one gets

$$\frac{\partial u_e}{\partial t} = -\nabla^T\cdot\left(\vec{E}\times\vec{H}\right) + \frac{1}{2}\left(\vec{D}^T\cdot\frac{\partial\vec{E}}{\partial t} - \vec{E}^T\cdot\frac{\partial\vec{D}}{\partial t}\right)$$
$$+ \frac{1}{2}\left(\vec{B}^T\cdot\frac{\partial\vec{H}}{\partial t} - \vec{H}\cdot\frac{\partial\vec{B}}{\partial t}\right) - \vec{E}^T\cdot\vec{j} \quad . \tag{2.2}$$

Coherent Excitations in Biological Systems
Ed. by H. Fröhlich and F. Kremer

This is in general no conservation law for the field energy and therefore an energy principle cannot be used to calculate body forces in media with losses. This is possible only when the dissipated power

$$Q = \vec{E}^T\cdot\vec{j} - \frac{1}{2}\left(\vec{D}^T\cdot\frac{\partial\vec{E}}{\partial t} - \vec{E}^T\cdot\frac{\partial\vec{D}}{\partial t} + \vec{B}^T\cdot\frac{\partial\vec{H}}{\partial t} - \vec{H}^T\cdot\frac{\partial\vec{B}}{\partial t}\right) \qquad (2.3)$$

becomes zero. For periodic fields of the frequency ω the time average of Q is of the form (we assume time dependence $\sim e^{-i\omega t}$ and $\vec{D} = \varepsilon\vec{E}$, $\vec{B} = \mu\vec{H}$)

$$\bar{Q} = \frac{1}{2}\left(\sigma' + \omega\varepsilon''\right)|\vec{E}_o|^2 + \frac{1}{2}\omega\mu''\,|\vec{H}_o|^2 \qquad (2.4)$$

where

$$\varepsilon'' = \mathrm{Im}\{\varepsilon\} \quad ; \quad \mu'' = \mathrm{Im}\{\mu\} \quad ; \quad \sigma' = \mathrm{Re}\{\sigma\} \quad .$$

For dissipative media or particles $(\varepsilon'' \neq 0$ or $\mu'' \neq 0$ or $\sigma' \neq 0)$ $\bar{Q}$ will be unequal zero. Therefore, the force calculation from the variation of the field energy must lead to meaningless results.

On the other hand there exists the electromagnetic momentum balance which can be used to derive meaningful expressions for the body force. For the local time change of $\vec{p}_e$ one gets with the help of Maxwell's equation

$$\frac{\partial}{\partial t}\vec{p}_e = (\Pi\cdot\nabla) - \vec{f}_e \qquad (2.5)$$

where Π is the Maxwell stress tensor and $\vec{f}_e$ the electromagnetic body force. There are infinite many ways to write the right hand side of (2.5) in the form of a balance equation. One must observe that the definition of Π determines $\vec{f}_e$. With that in mind one makes the following choice for Π

$$\Pi = \frac{1}{2}\left(\vec{E}\vec{D}^T + \vec{D}\vec{E}^T + \vec{H}\vec{B}^T + \vec{B}\vec{H}^T - \left(\vec{D}^T\cdot\vec{E} + \vec{B}^T\cdot\vec{H}\right)U\right) \qquad (2.6)$$

which is a symmetric tensor. U means the unit tensor, products with a dot are scalar products and without a dot dyadic products of vectors. With (2.6) for Π one gets for the body force

$$\begin{aligned}\vec{f}_e = \frac{1}{2}\Big(&(\nabla^T\cdot\vec{E})\vec{D} - (\nabla^T\cdot\vec{D})\vec{E} + \vec{D}\times\nabla\times\vec{E} - \vec{E}\times\nabla\times\vec{D} \\ &+ (\nabla^T\cdot\vec{H})\vec{B} + \vec{B}\times\nabla\times\vec{H} - \vec{H}\times\nabla\times\vec{B}\Big) \qquad (2.7) \\ &+ \rho\vec{E} + \vec{j}\times\vec{B} + \frac{\partial}{\partial t}\left(\vec{D}\times\vec{B} - \frac{1}{c^2}\vec{E}\times\vec{H}\right) \quad .\end{aligned}$$

This expression contains the well known terms of forces on charges and currents. The additional terms are related to dielectric and magnetic inhomogeneities and the radiation pressure. For the vacuum $\vec{f}_e$ becomes zero.

For the total momentum, the sum of mechanical momentum $\tau\vec{v}$ and $\vec{p}_e$ there exists a conservation law (in the nonrelativistic approximation)

$$\frac{\partial}{\partial t}\left(\tau\vec{v} + \vec{p}_e\right) = \left(-\tau\vec{v}\vec{v}^T + \Pi - P\right)\cdot\nabla \qquad (2.8)$$

if no other external forces are present. P is the pressure tensor. Only the difference Π - P has a physical meaning. Combination of (2.5) and (2.8) leads to a material momentum balance.

$$\frac{\partial}{\partial t}\left(\tau\vec{v}\right) = -\left(\tau\vec{v}\vec{v}^T + P\right)\cdot\nabla + \vec{f}_e \tag{2.9}$$

where the pressure tensor P depends on the field quantities and the thermodynamic state. This dependence is determined by the choice of $\vec{f}_e$ and is not yet specified. If the electromagnetic field becomes zero, P goes over into the usual mechanical pressure tensor.

For periodic fields $\vec{E}(\vec{r}, t) = \mathrm{Re}\{\vec{E}(\vec{r})e^{-i\omega t}\}$ of the frequency ω one gets for the time average of the body force

$$\overline{\vec{f}_e} = -\frac{1}{4}\mathrm{Re}\{|\vec{E}|^2\nabla\varepsilon + |\vec{H}|^2\nabla\mu - \varepsilon^*\vec{E}\vec{E}^{T*}\,\nabla\left(\frac{\varepsilon}{\varepsilon^*}\right) - \mu^*\vec{H}\vec{H}^{T*}\,\nabla\left(\frac{\mu}{\mu^*}\right) + i\omega\left(\vec{E}^*\times\vec{H} + \vec{E}\times\vec{H}^*\right)\,\varepsilon\mu\} \quad . \tag{2.10}$$

Here we assumed $\rho = 0$; $\vec{j} = 0$; $\vec{D} = \varepsilon\vec{E}$; $\vec{B} = \mu\vec{H}$ with complex and frequency dependent ε and μ.

Use has been made of Maxwell's equations for periodic fields in the complex notation

$$\begin{aligned} \nabla\times\vec{H} &= \vec{j} - i\omega\vec{D} \quad ; \quad & \nabla\times\vec{E} &= i\omega\vec{B} \\ \nabla\cdot\vec{D} &= \rho \quad ; \quad & \nabla^T\cdot\vec{B} &= 0 \end{aligned} \tag{2.11}$$

* means the complex conjugated quantity. In (2.10) there we have terms proportional to the gradient of ε and μ well known from the body force in electro and magneto statics. For media with losses we get additional terms which depend on the gradients of $\varepsilon/\varepsilon^*$ and μ/μ^* . Furthermore, there are terms with $\vec{E}^*\times\vec{H}$, which contribute to the body force at higher frequencies.

In the time average the electromagnetic momentum balance equation goes over into

$$(\overline{\Pi}\cdot\nabla) = \overline{\vec{f}_e} \tag{2.12}$$

with

$$\overline{\Pi} = \frac{1}{4}\mathrm{Re}\,\{\varepsilon\}\cdot\left(\vec{E}^*\vec{E}^T + \vec{E}\vec{E}^{T*} - |\vec{E}|^2\,U\right) + \frac{1}{4}\mathrm{Re}\,\{\mu\}\left(\vec{H}^*\vec{H}^T + \vec{H}\vec{H}^{T*} - |\vec{H}|^2\,U\right) \quad . \tag{2.13}$$

For the material momentum balance one gets in the time average, if no other external forces are present

$$(\overline{P}\cdot\nabla) = \overline{\vec{f}_e} \quad . \tag{2.14}$$

Here we neglect the convective contribution $\tau\vec{v}\vec{v}^T$ as small at not too high field strength.

With an external force $\vec{f}_{ex}$, which keeps the body at rest, we get

$$\overline{P}\cdot\nabla - \overline{\vec{f}_e} = \vec{f}_{ex} \qquad (2.15)$$

and because of (2.12)

$$\int \vec{f}_{ex}\, dv_s = \int (\overline{P}_s - \overline{\Pi}_s)\cdot\vec{n}_s\, da_s \qquad (2.16)$$

where v_s is the volume of a solid body, a_s its surface and $\vec{n}_s$ the surface normal vector. Use has been made of Gauss' theorem.

At a surface of discontinuity between a solid body (s) and a liquid (l) we get a surface force $\vec{t}_{sl}$ in general. For a surface at rest this surface force must be compensated by an external force and we have

$$\vec{t}_{ex} = -\vec{t}_{sl} \quad . \qquad (2.17)$$

From limit considerations of (2.12) and (2.14) applied on a small volume including the surface of discontinuity one gets

$$\vec{t}_{sl} = (\overline{\Pi}_l - \overline{P}_l - \overline{\Pi}_s + \overline{P}_s)\cdot\vec{n}_s \quad . \qquad (2.18)$$

The total external force, which keeps the body at rest in the time average is given by

$$\vec{F}_{ex} = \int \vec{t}_{ex}\, da_s + \int \vec{f}_{ex}\, dv_s \quad . \qquad (2.19)$$

With the help of the equations (2.16) - (2.18) one gets.

$$\vec{F}_{ex} = \int (\overline{P}_l - \overline{\Pi}_l)\cdot\vec{n}_s\, da_s \quad . \qquad (2.20)$$

Equation (2.20) enables one to calculate the force acting on the center of mass of a solid body immersed in a liquid if the pressure tensor and the Maxwell stress tensor are known in the liquid at the surface of the body. An external force $\vec{F}_{ex}$ prevents the movement of the center of mass. Similar considerations allow to calculate the external torque, which prevents the body from rotation. This will not be done here, because we deal finally with homogeneous rigid spheres where no rotation should occur in a homogeneous field.

In the following we discuss (2.20) in special cases. For this we have to solve the Maxwell equations for special boundary conditions and take into account the material momentum balance (2.14) for the liquid. Because the general problem is too complex, we cannot give a general solution of the problem. Many approximations must be done to arrive finally at applicable formulas.

3. Force on a Body Immersed in a Fluid. Spherical Particle with Dielectric Losses

We start with equation (2.20) for the calculation of the force on a spherical particle in an external RF field (RF = radio frequency). Under the conditions that the radius a of the particle is small compared to the distance d of the electrodes and this distance is small compared to the vacuum wave length $\lambda_o = \frac{\omega}{2\pi c}$ we can neglect the terms with the magnetic field strength $\vec{H}$ in the Maxwell stress tensor (2.13) and the body force (2.10). This terms are small of the order $\frac{d}{\lambda_o}$. The Maxwell equations go over into

$$\nabla\times\vec{E} = 0 \quad \text{or} \quad \vec{E} = -\nabla\Phi \quad . \tag{3.1}$$

This means for the field calculation we have an electrostatic problem.
For the determination of the pressure tensor $\bar{P}$ we have to solve (2.14) with the given expression for $\vec{f}_e$. In the special case we discuss here, (2.14) goes over into

$$(\bar{P}\cdot\nabla) = -\frac{1}{4}\,\mathrm{Re}\,\{\,|\vec{E}|^2\nabla\varepsilon - \varepsilon^*\,\vec{E}\vec{E}^{T*}\,\nabla\left(\frac{\varepsilon}{\varepsilon^*}\right)\,\} \quad . \tag{3.2}$$

In general, equation (3.2) does not lead to hydrostatic equilibrium with $\bar{P} = pU$ where p is a scalar pressure. Therefore one can expect the appearance of a hydrodynamic motion in media with dielectric losses at higher field strength.
For lower field strength we neglect the contributions of the terms with $\nabla\varepsilon$ and $\nabla\,{}^{\varepsilon}/\varepsilon^*$ and get a constant pressure tensor P if there are no other sources for a hydrodynamic motion. A constant pressure tensor does not contribute to the surface integral (2.20) and we get for the external force, which keeps the particle at rest

$$\vec{F}_{ex} = -\frac{1}{4}\,\mathrm{Re}\,\{\varepsilon_1\}\int\left(\vec{E}_1^*\vec{E}_1^T + \vec{E}_1\vec{E}_1^{T*} - |\vec{E}_1|^2\,U\right)\cdot\vec{n}_s\,da_s \quad . \tag{3.3}$$

If the particle is rigid and homogeneous, the surface integral (3.3) can be transformed into

$$\vec{F}_{ex} = -\frac{1}{4}\,\mathrm{Re}\,\{\varepsilon_1\}\int\Big[(b^* - b)\left(\vec{E}_s^{T*}\cdot\vec{n}_s\right)\vec{E}_s + b|\vec{E}_s|^2\,\vec{n}_s + |b|^2\,|\vec{E}_s^T\cdot\vec{n}_s|^2\,\vec{n}_s\Big]\,da_s \tag{3.4}$$

with

$$b = \frac{\varepsilon_s - \varepsilon_1}{\varepsilon_1} \quad . \tag{3.5}$$

The surface integral (3.4) contains only field quantities at the solid side of the liquid-solid interface. Use has been made of the transition conditions

$$\varepsilon_1\vec{E}_1^T\cdot\vec{n}_s = \varepsilon_s\vec{E}_s^T\cdot\vec{n}_s \quad \text{and} \quad \vec{E}_1\times\vec{n}_s = \vec{E}_s\times\vec{n}_s$$

at the solid-liquid interface and of the fact that we assumed $\nabla\varepsilon_s = 0$ (homogeneous particle).

If one applies (3.4) for a special case, various approximations must be made to evaluate the surface integral in equation (3.4). For a homogeneous rigid sphere in a slightly inhomogeneous field we make use of the gradient approximation. This means, we express the field at the surface by a Taylor expansion around the center of the sphere up to first order terms. This expansion reads

$$\vec{E}_s(a\,\vec{n}_s) = \vec{E}_s(0) + a\,\left(\vec{E}_s\nabla^T\right)_o \cdot \vec{n}_s \tag{3.6}$$

where $\vec{E}(0)$ is the electric field strength at the center of the sphere, $(\vec{E}_s\nabla^T)_o$ the gradient of $\vec{E}$ at the center and a the radius of the sphere. For the integral (3.4) one gets

$$\vec{F}_{ex} = -\frac{V}{20}\left[b^*(2b+5)\left(\nabla\vec{E}^T\right)\cdot\vec{E}^* + b\left(2b^*+5\right)\left(\nabla\vec{E}^{T*}\right)\cdot\vec{E}\right]\cdot \mathrm{Re}\,\{\varepsilon_1\} \tag{3.7}$$

V is the volume of the sphere. The field and the gradient must be taken at the center of the sphere. A similar integration has been performed by Denner (6) in the electrostatic case.

One can now introduce the external field $\vec{E}_o$ instead of the actual field $\vec{E}$ inside the sphere. The external field is the field that exists at the point $\vec{r}$ inside the sphere if $\varepsilon_s = \varepsilon_1$. In general, there is a functional relation between the actual field and the external field. At the center of the sphere one gets

$$\vec{E} = \frac{3}{b+3}\vec{E}_o \tag{3.8}$$

and

$$\nabla\vec{E}^T = \frac{5}{2b+5}\,\nabla\vec{E}_o^T \quad . \tag{3.9}$$

These relations can be calculated from a multipole expansion. They are valid only at the center of the sphere. For the external force one gets

$$\vec{F}_{ex} = -\frac{3}{16}\,V\left(\frac{b^*}{3+b^*} + \frac{b}{3+b}\right)\left(\varepsilon_1^* + \varepsilon_1\right)\,\nabla|\vec{E}_o|^2 \quad . \tag{3.10}$$

This expression is valid for slightly inhomogeneous fields. It goes over into the well known result of electrostatics in the case $b^* = b$ and $\varepsilon_1 = \varepsilon_1^*$ when one observes that $\vec{E}_o$ is the amplitude of the external field.

We compare equation (3.10) with a formula for the external force derived by Sher (7) from an energy principle. Sher's result is given by

$$\vec{F}_{ex}(\mathrm{Sher}) = -\frac{3}{8}\,V\left(\frac{\varepsilon_1^*\,b}{b+3} + \frac{\varepsilon_1\,b^*}{b^*+3}\right)\nabla|\vec{E}_o|^2 \tag{3.11}$$

which is different from (3.10) for complex dielectric constants ε_1 and ε_s. The difference between the two expressions has the form

$$\vec{F}_{ex} - \vec{F}_{ex}(\text{Sher}) = -\frac{9}{16} V \frac{(b-b^*)(\varepsilon_1-\varepsilon_1^*)}{|3+b|^2} \nabla|\vec{E}_o|^2$$

$$= \frac{9}{4} V \frac{(\varepsilon_s'' \varepsilon_1' - \varepsilon_s' \varepsilon_1'')\varepsilon_1''}{|\varepsilon_1|^2 \; |3+b|^2} \nabla|\vec{E}_o|^2 \quad . \tag{3.12}$$

Here we have introduced $\varepsilon_1 = \varepsilon_1' + i\,\varepsilon_1''$ and $\varepsilon_s = \varepsilon_s' + i\,\varepsilon_s''$. The difference becomes zero if $\varepsilon_1'' = 0$ or if $\varepsilon_s'/\varepsilon_s'' = \varepsilon_1'/\varepsilon_1''$. This means the loss angle of liquid and solid must be the same. For $\varepsilon_s'' = 0$ the difference between the force expressions becomes proportional to $\varepsilon_1''^2$. So we come to the conclusion that Sher's formula cannot be applied for media with large losses when the deviations (3.12) become large. This is the case for conducting media at low frequencies where the imaginary part of ε_1 becomes equal to σ/ω with σ being the conductivity.

The gradient approximation breaks down for stronger inhomogeneous fields. In this case one must go back to the original expression (2.20) and find other ways to evaluate the surface integral. This situation occurs for example when another particle is in the close vicinity of the particle of interest.

4. *Force between Two Spherical Particles in a Homogeneous External Field.*

We consider two identical, spherical particles in a homogeneous external field $\vec{E}_o$ in the x-direction. The particles may have arbitrary positions with respect to the field. The evaluation of the integral (2.20)(Sauer (8)) leads to an expression for the resulting force on particle number 1 in the time average

$$\vec{F}_1 = -\frac{\alpha E_o^2}{\xi^4} \left[2\xi_{ox}\left(1 + \frac{s\,a^3}{\xi^3}\right)\vec{e}_x + \left[\left(1 - 5\xi_{ox}^2\right) - \frac{2s\,a^3}{\xi^3}\left(1 + 4\xi_{ox}^2\right)\right]\vec{\xi}_o \right] + O\!\left(\frac{a}{\xi^9}\right) \tag{4.1}$$

with

$$\alpha = \frac{27\,V^2\,|b|^2\,\mathrm{Re}\{\varepsilon_1\}}{8\pi\,|b+3|^2} \quad ; \quad b = \frac{\varepsilon_s - \varepsilon_1}{\varepsilon_1} \tag{4.2}$$

and

$$s = \mathrm{Re}\left\{\frac{b}{b+3}\right\} \tag{4.3}$$

where V is the volume of the sphere of radius a, ξ is the distance of the centers of the two particles, $\vec{\xi}_o$ is the unit vector in direction of the connection line between the two particles and $\vec{e}_x$ the unit vector in the x-direction. ξ_{ox} is equal to cosine of the angle between $\vec{e}_x$ and $\vec{\xi}_o$. Use has been made of a series solution for the field (Sauer (9)) with the expansion parameter $^a/\xi$. In equation (4.1) terms of the order $(^a/\xi)^9$ are neglected. The force vector $\vec{F}_1$ lies in the plane defined by $\vec{e}_x$ and $\vec{\xi}_o$.

For $\vec{\xi}_o$ parallel to $\vec{e}_x$ the force between two identical particles is attractive and for $\vec{\xi}_o$ perpendicular to $\vec{e}_x$ the force is repulsive. For dielectric particles with $a = 10^{-4}$cm suspended in water at a field strength of 700 V/cm (amplitude) and $\xi = 4a$ the attractive force (particles parallel to the field) becomes approximately

$$F_1 \approx 2 \cdot 10^{-13} [\text{ Newton }] \quad .$$

The terms, proportional to s in equation (4.1), are minor corrections for distances $\xi > 4a$. The force $\vec{F}_1$ can be derived from a potential U. We have

$$\vec{F}_1 = - \nabla_\xi U \tag{4.4}$$

with

$$U = \frac{2\alpha E_o^2}{3} \left(\frac{P_2(\cos\theta)}{\xi^3} + \frac{s\, a^3}{\xi^6} \left[P_2(\cos\theta) + 2 \right] \right) + O\left(\left(\frac{a}{\xi}\right)^8 \right) \tag{4.5}$$

neglecting terms of the order $\left(\frac{a}{\xi}\right)^8$. P_2 is the Legendre polynomial of second degree and θ the angle between $\vec{\xi}_o$ and $\vec{e}_x$. The first term in equation (4.5) was allready proposed by Saito (2) from the consideration of interacting dipoles. Saito does not give an expression for α. Actually, one can look on the terms without s in (4.1) as the force of interaction between two induced dipoles. The correction terms with s go beyond a dipole approximation. For spheres at closer distances one has to take into account higher multipoles.

The force F_1 is a potential force due to the fact that the approximation we used coincides with the gradient approximation (3.6). One observes that $\nabla^T \cdot \vec{F}_1$ is unequal zero. The derivation of $\vec{F}_1$ from a potential U is formally and does not reflect any relation to the interpretation of the field energy in electrostatics.

5. Trajectories of Particles in the Process of Pearl Chain Formation.

For higher electric field strength one can neglect the influence of Brownian motion on the particle movement. If $U >> kT$ (k is the Boltzmann constant and T the absolute temperature) the equation of motion for the relative movement of the particles reads

$$6\pi\, \eta_1\, a \frac{d\vec{\xi}}{dt} = - 2\, \vec{F}_1 \quad , \tag{5.1}$$

with

$$\vec{\xi} \equiv \xi\, \vec{\xi}_o = \vec{r}_2 - \vec{r}_1 \tag{5.2}$$

and η_1 is the viscosity of the liquid. Here we assumed that the Reynolds number is small compared to one, and therefore we neglect the acceleration term. This is justified for most of the experiments. Equation (5.1) is valid for larger distances because we assumed Stokes law for the hydrodynamic resistance of the particles. From the expression (4.1) for the force one concludes that the motion of the particles is confined to the plane defined by $\vec{e}_x$ and $\vec{\xi}_o$ at $t = 0$. So we have a two dimensional motion only.

In special cases (5.1) can be solved in a closed form. Introduction of the dimension less quantities

$$\tau = \frac{2\alpha E_o^2 t}{6\pi\eta_1 a^6} \qquad \text{and} \qquad \vec{\gamma} = \frac{\vec{\xi}}{a} \tag{5.3}$$

leads to the system of differential equations

$$\frac{d\vec{\gamma}}{dt} = \frac{1}{\gamma^4}\left[2\gamma_{ox}\left(1 + \frac{s}{\gamma^3}\right)\vec{e}_x + \left[\left(1 - 5\gamma_{ox}^2\right) - \frac{2s}{\gamma^3}\left(1 + 4\gamma_{ox}^2\right)\right]\vec{\xi}_o \right] \tag{5.4}$$

with

$$\gamma_{ox} = \xi_{ox} \quad , \quad \gamma_{oy} = \xi_{oy} \quad , \quad \gamma_{oz} = \xi_{oz} \quad .$$

For the initial conditions

$$\gamma_{oy}(0) = 0 \quad ; \quad \gamma_{ox}^2 = 1 - \gamma_{oz}^2 = \gamma_{ox}^2(0) \quad ; \quad \gamma = \gamma(0) \quad \text{at} \quad \tau = 0 \tag{5.5}$$

equations (5.4) can be integrated in a closed form, if $s \ll 1$. One gets then for $\gamma = \xi/a$

$$\frac{\gamma(\tau)}{\gamma(0)} = \left[\frac{\gamma_{ox}(\tau)\left(1 - \gamma_{ox}^2(\tau)\right)}{\gamma_{ox}(0)\left(1 - \gamma_{ox}^2(0)\right)}\right]^{1/2} \quad . \tag{5.6}$$

This is a parameter representation of the trajectories in the x - z plane. If the initial angle θ_o is larger than 63.4^o, the curved trajectory (5.6) goes parallel to x-axis at $\theta = 63.4^o$, reaches a maximum distance between the particles at $\theta = 54.7^o$ is parallel to the z-axis at an angle $\theta = 39.2^o$ as long as $a/\xi \ll \frac{1}{2}$.

For the time τ one gets

$$\tau = \frac{1}{2\left(\gamma(0)\right)^5\left[\gamma_{ox}(0)\left(1 - \gamma_{ox}(0)^2\right)\right]^{5/2}} \int_{\gamma_{ox}(0)}^{\gamma_{ox}(\tau)} \left[x\left(1 - x^2\right)\right]^{3/2} dx \tag{5.7}$$

where the integral has the value

$$\int \left[x\left(1 - x^2\right)\right]^{3/2} = \frac{2}{77}\left[\left(-7x^4 + 13x^2 - 4\right)y - 2\sqrt{2}\, F\left(\arccos\sqrt{x}\, , \frac{1}{\sqrt{2}}\right)\right]$$

with $y = x(1 - x^2)$. $F(u, k)$ is the incomplete elliptic integral of the first kind.

In the case of a direct approach of the particles ($\vec{\xi}$ parallel to $\vec{e}_x$) the solution is given by

$$\tau = \frac{1}{10}\left(\gamma^5(0) - \gamma^5(\tau)\right) + s\,\gamma^2(\tau) - \gamma^2(0)\big) - 8s^2 \int_{\gamma(0)}^{\gamma(\tau)} \frac{\gamma d\gamma}{\gamma^3 + 4s} \quad . \qquad (5.8)$$

The value of the integral can be found in the tables.
The results of this paragraph, especially equations (5.6), (5.7) and (5.8), can be used for the quantitative analysis of the process of pearl chain formation in dilute suspensions. At lower electric field strength the effects of Brownian motion must be taken into account.

6. Summary and Conclusions.

Interaction forces between suspended dielectric particles in an external high frequency field were calculated with the help of the electromagnetic momentum balance. If the medium and the particles have dielectric losses, the results of the calculation differ from the results one gets by using an energy principle (Sher (7), Pohl (10)). Because the electromagnetic field energy is not conserved, the application of an energy principle leads to wrong results with respect to the dependence of the forces on the dielectric properties of the medium. Furthermore, the influence of electric double layers on the forces in electrolyte solutions at lower frequencies must be reinvestigated.
Equation (4.1) for the interaction force can be generalized when the particles have different volumes and different dielectric properties. Then α becomes proportional to $V_1 \cdot V_2$ instead of V^2. Particles with different sign of $\varepsilon_s - \varepsilon_1$ show repulsion for allignement parallel to the field and attraction perpendicular to the field.

Acknowledgements

I thank Professor H.P.Schwan for bringing my attention to this problem and for many valuable discussions. The contributions of Professor R. Schlögl especially to the calculations of the field distribution were a great help.

List of Symbols

$\vec{D}$ = vector of dielectric displacement
$\vec{E}$ = vector of electric field strength
$\vec{B}$ = vector of magnetic induction
$\vec{H}$ = vector of magnetic field strength
ρ = electric charge density
$\vec{j}$ = vector of conductive current density
τ = mass density
c = phase velocity of light in vacuum
$\vec{v}$ = vector of center of mass velocity
ε = dielectric constant
μ = magnetic permittivity

References

1. Krasny-Ergen W., Hochfreq. u. Elektroak., 48, 126 (1936).
2. Saito M., and H.P. Schwan in: Biological Effects of Microwave Radiation Mary Fouse Peyton Ed., page 85-97, Plenum Press, New York 1961.
3. Landau L.D., and E.M. Lifshitz "Electrodynamics of Continuous Media". Course of Theoretical Physics Vol.8, page 253-256, Pergamon Press, New York 1960.
4. Meixner J., Z.f.Physik 229, 352-364 (1969).
5. De Groot S.R., P.Mazur "Nonequilibrium Thermodynamics", page 376-404 North Holland, Amsterdam 1962.
6. Denner V., H.A. Pohl: Research Note 127, Quantum Theoretical Research Group, Oklahoma State University, Stillwater, OK (1982).
7. Sher L.D., Nature 220, 695-696 (1968)
8. Sauer F.A.: "Ponderomotive Forces on Dielectric Materials with Losses in Periodic Fields", to be published
9. Sauer F.A. and R. Schlögl: "Trajectories of Long Cylinders and Spheres in a Homogeneous High Frequency Field" to be published
10. Pohl H.A., "Dielectrophoresis", Cambridge University Press, 1978.

Coherent Excitations in Blood

S. ROWLANDS
Faculty of Medicine, University of Calgary, Calgary, Alberta, Canada, T2N 1N4

I. Introduction. In Calgary, W.G. deHaas has had spectacular success in treating fractured bones which would not unite and infected bones which would not heal (deHaas, Lazarovici & Morrison 1979; deHaas, Morrison & Watson 1980) with pulsed magnetic fields. Patients have been spared otherwise inevitable amputations. Dr. deHaas' laboratory is next to mine and I always had an interest in his work. Two years ago while on Sabbatical leave I began to look for an explanation of the induced healing process at the cellular level. In the course of the search I came across the Fröhlich (1968,1980) theory of coherent excitations in living cells. I have not yet explained

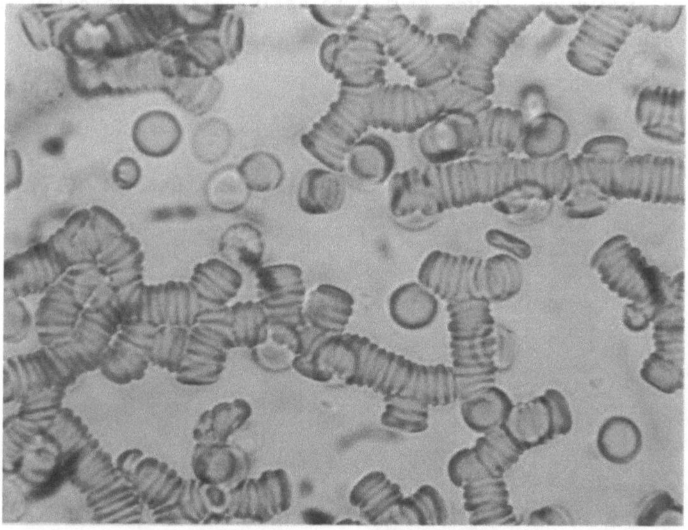

Fig.1. Rouleau formation in normal blood

Coherent Excitations in Biolcgical Systems
Ed. by H. Fröhlich and F. Kremer

bone healing in terms of the theory but it has given me an explanation of observations which I made years ago in the course of experiments on what is called rouleau-formatior in blood (Fåhraeus 1929; Rowlands & Skibo 1972; Kernick *et al.* 1973). When blood is at rest the erythrocytes (red blood cells, RBC), which are discoids of diameter 8 µm, array themselves face to face and give the appearance under the microscope of stacks of coins (rouleaux, Fig.1). Rouleau formation is increased in a number of diseases and it is assessed in the clinic by the erythrocyte sedimentation rate. The adhesion of the cells is brought about by the macromolecules in plasma (the fluid phase of blood) and an increased concentration of macromolecules in disease enhances the adhesion.

II. Initial Observations. Blood cells are slightly more dense than plasma and so they settle. They can, therefore, be viewed in a plane in the living state under a microscope. Red cells are not motile and their only motion is a slow Brownian movement which can scarcely be detected directly but which we study by taking a picture every 10 to 20 sec and then projecting the film at ordinary ciné-projection rates. For analysis, measurements are made frame by frame on this film. I noticed years ago that when two red cells came close they seemed to "rush" together on the time scale of the slow Brownian movement. At the time I had no explanation and it could have been an illusion. So there was no incentive to authenticate and to measure the phenomenon until the Fröhlich theory came along.

The other observation was on rouleaux themselves. I have long thought that rouleau-formation is a minor but essential step in the complex mechanisms of the clotting of blood and that abnormal clotting, for instance, in coronary heart disease, might be associated with abnormal rouleaux. I was trying to estimate the intercellular adhesive force by traction on each end of a rouleau. What was intriguing and what

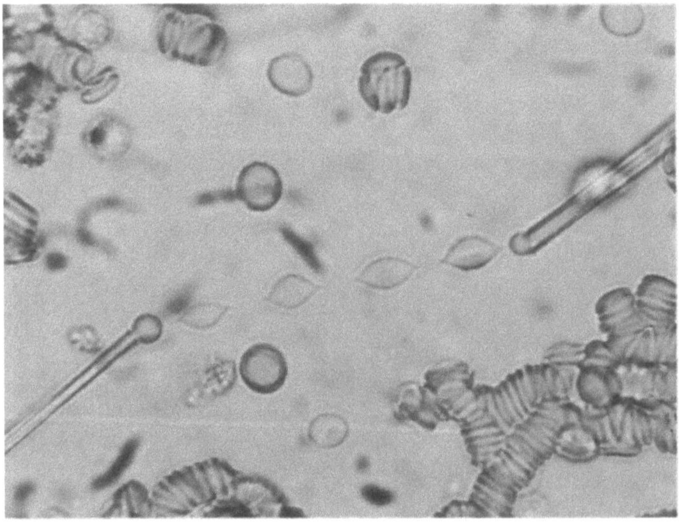

Fig. 2.
(Legend see p. 147)

turned out to be relevant to the Fröhlich phenomena was this: the cells came away from each other but remained attached by scarcely visible threads (Fig.2) which could reach lengths of ten or more times the cellular diameter. Moreover, if the tractive force was released, the threads rapidly pulled the string of cells together, right back into a rouleau. This seemed to be an active (energy-requiring) process because, had the threads been merely viscoelastic, there would have been a rebound, but not down to zero length.

III. Deviations from Brownian Movement. It was first necessary to prove that interacting erythrocytes deviated from Brownian movement. Secondly it was necessary to demonstrate that the interaction behaved in a manner compatible with the postulates of Fröhlich's theory. The experiments have been published in detail (Rowlands *et al.* 1981; Rowlands, Sewchand & Enns 1982a,b). The cells lie on a ruled, optically-flat, horizontal surface and so their movement can be accurately measured with standard time-lapse ciné-microphotography. When two cells touch they adhere and quickly (on the time-scale of the experiments) align themselves into a two-cell rouleau. Adhesion and alignment also occur when a cell touches a pre-existing rouleau or when two rouleaux collide (Figs.3a,b). The process is therefore similar to the coagulation of a colloid but with much larger "particles" and with aggregates which are more regular. Smoluchowski (1917) developed a theory of colloid aggregation resulting from Brownian motion in three dimensions and we have adapted this theory to the movement of cells in a plane. The Smoluchowski theory of colloid aggregation needs further modification before it can be used to monitor deviations from random motion for the following reason. In the theory it is assumed that the diffusion coefficient of the individual particles is given by the Stokes-Einstein relation

(1) $$D = \frac{kT}{6\pi\eta r}$$

where the denominator is the Stokes' drag coefficient (μ) for rigid spheres of radius r in a medium of viscosity η

(2) $$\mu = 6\pi\eta r$$

The drag coefficient gives the relationship between the force F on a rigid sphere and its terminal velocity v in a fluid of viscosity η

(3) $$F = \mu v$$

Fig.2. The end cells of a six-cell rouleau have been gently aspirated into the glass micropipettes seen left and right. The pipettes were then drawn apart. The contractils are not resolved by light microscopy. In this picture their presence is inferred by the distortion of the shape of the cells in the chain. On release of negative pressure in the pipettes a normal-looking six-cell rouleau reforms

Erythrocytes are discoids and not spheres and they are far from being rigid. Perrin gave a correction to Stokes' equation for spheroidal particles and if this is applied to RBC there is reasonable agreement with experimental observations (Canham, Jay & Tilsworth 1971) on the free fall of erythrocytes in liquids. The small discrepancy can be accounted for by the flexibility of the cells (Rowlands *et al.* 1982b).

But there is another discrepancy which arises inevitably from the design of the experiments. We measured the Brownian movement of isolated red cells resting on a plane surface (Sewchand, Rowlands & Lovlin 1982) and calculated the drag coefficient μ from Einstein's (1926) relation for the mean square deviation $\langle U^2 \rangle$ of particles undergoing Brownian motion as a function of time, t:

$$\langle U^2 \rangle = 4kTt\mu^{-1} \quad (4)$$

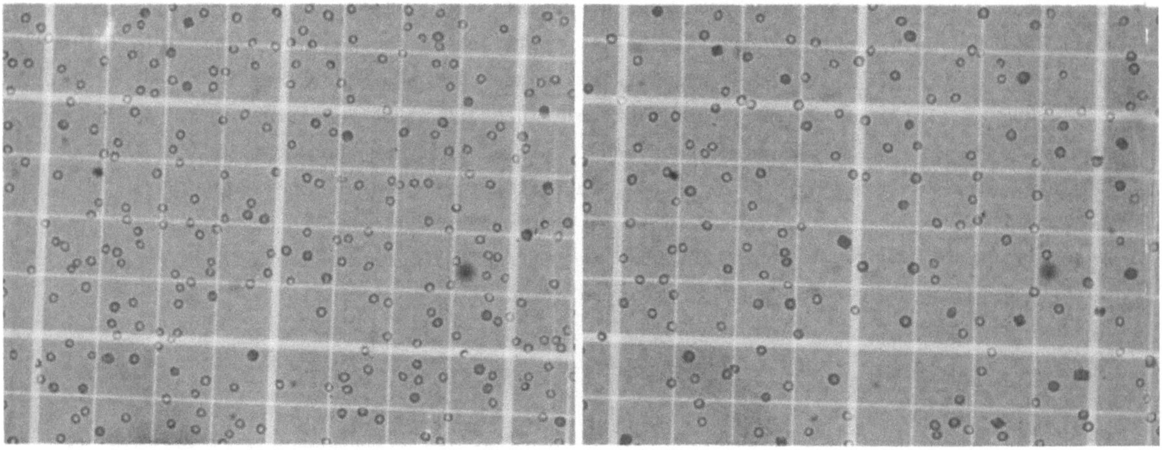

Fig.3a(left) Distribution of the erythrocytes on the floor of the haemacytometer chamber at the beginning of an experiment. The darker rings are small rouleaux seen on end. There are also two prominent fuzzy spots which are artefacts

Fig.3b(right) The same field of view about an hour later. Note the diminution in the total number of rings and the presence of larger rouleaux (pseudo-rectangles) which have fallen onto their sides

and found that the drag μ was consistently higher than that given by equation (2) even with Perrin's correction. We also found the same proportional increase in drag for rigid microspheres. It is clear that cells and particles, slightly more dense than the surrounding fluid, and thus resting on the lower surface of their enclosure, experience an increased hydrodynamic drag caused by proximity to the surface. Fortunately we were able to coat the surface with silicone in a way that made this increased drag consistent, even when the cells were modified to test the postulates of Fröhlich's theory. Over the series of experiments the standard error of the mean of the drag coefficient was less than 4% (Rowlands *et al.* 1982b). Nevertheless, because the drag coefficient is a measure of unperturbed Brownian motion, against which we can observe the deviations from Brownian motion, it is *measured in the course of each experiment* by observations on cells which do not come close enough to other cells to interact. From this experimental measurement the true value of the diffusion coefficient (D equation (1)) is calculated and incorporated in the Smoluchowski equations. When this is done the microspheres "coagulate" at the rate they should if the collisions are simply the result of Brownian motion (Rowlands *et al.* 1982b).

When we do the same experiments on living human erythrocytes in their own plasma the rate of coagulation is three times that expected for Brownian coagulation. The difference is statistically significant ($p<0.001$). The Smoluchowski theory of colloid aggregation incorporates the possibility of an attraction between the particles and permits the calculation of the range of this attraction. A tripling of the coagulation rate implies an attraction at a centre to centre distance of three cell radii. The cell radius is 4 μm. Therefore the attraction begins to be effective when the cell membranes are 4 μm apart (Rowlands *et al.* 1982b) a distance of several orders of magnitude greater than the range of chemical forces.

We therefore had an authentic ultra-long-range interaction and the next step was to test if it had features in common with the predictions of the Fröhlich theory of coherent excitations in cells. This theory (Fröhlich 1968,1980) predicts that there should be interactions at a range of several micrometres between cells and between macromolecules and organelles within and without cells, driven by a phonon-maser-like process in living membranes. The major requirements for a Fröhlich process are

(a) an organized array of molecules as in a biological membrane
(b) a membrane potential to polarize the molecules in the membrane
(c) a supply of energy to the membrane, which if it is above a minimum value, will trigger membrane vibrations into a coherent mode

We carried out the following tests (Rowlands *et al.* 1981,1982a,b):

(a) Red cells were treated with glutaraldehyde which cross-links proteins and amino phospholipids and so disorganizes the membrane. Some enzyme and receptor activity remains and, at the concentration of glutaraldehyde which we used, the shape, area and volume of the cells was unchanged. Suspended in their own plasma these cells showed no deviation from Brownian movement and so no interaction.
(b) The red cell membrane potential was reduced to zero by bringing the pH of the plasma from 7.6 to 6.3. This abolished the interaction and it was restored in the same cells by returning the pH to 7.6. Although it can be argued that this change in pH would have little effect on the properties of plasma we were able to use another method of abolishing the membrane potential. Treatment of the cells with ionophore A23187 makes the membrane leaky to ions and therefore unable to sustain a membrane potential. Again the long-range interaction disappeared.

(c) Finally cells were depleted of their source of metabolic energy by letting them use up their stores. This also abolished the interaction which returned when the same cells were recharged with their source of energy.

In these various ways we were able to show that the interaction between human erythrocytes behaved in accordance with the major postulates of Fröhlich's theory of coherent excitations in cells. All these results have recently been confirmed (O.G. Fritz, Jnr. private communication) by an independent method, quasi-elastic light scattering.

IV. Specificity of the Interaction. Fröhlich predicts that the interaction should be specific; attractive, repulsive or zero depending on the frequencies or phases of the oscillations of the interacting systems. This specificity offers explanations (Cooper 1981) for many currently inexplicable intracellular processes. It can also explain how, in the apparent chaos of the early embryo, like cells can recognize and join with each other. Experimentally a random mixture *in vitro* of initially isolated embryonic liver and heart cells will sort itself out into a recognizable pattern (Steinberg & Wiseman 1972) even when active cell movement is inhibited. There is some evidence of a specific erythrocyte interaction in the experiments of Sewchand & Canham (1976). They studied rouleau-formation among mixtures of cells from ten pairs of different mammalian species. Contrary to predictions from the coagulation of poly-dispersed colloids (Overbeek 1952) cells of the same species show a preference for each other. There was an exception in the mouse-rat combination of erythrocytes, but this exception makes it the more likely that the non-random pairing in the other cases was an active process (Rowlands *et al.* 1982a,b).

There is no interpretation of the experiments of Sewchand & Canham (1976) other than in terms of the Fröhlich phenomenon. In normal rouleau formation, when two cells touch, they adhere permanently and subsequently they slide together to form a rouleau. If, in the inter-species experiments, this did not always apply; if, when two cells, one from each species, touched there was a possibility of them coming apart, the result would be as Sewchand & Canham observed. To settle this point we have done a number of experiments using the "Brownian" method but with mixtures of either human and cat cells or of human and dog cells. Exhaustive examination of the ciné-films in fourteen experiments showed that no cells in contact ever came apart again. We have also made measurements of interaction coefficients in the cat-human experiments. These experiments cannot be done in plasma because of immunological reactions and we have yet to find an artificial medium which is equally suitable to both cat and human cells. But as a preliminary finding we have a significantly ($p<0.005$) lower interaction coefficient for the mixture of cells than for either of the pair measured alone (Sewchand & Rowlands 1983). It is therefore reasonable to conclude, on the basis of Sewchand & Canham's work, combined with the above-described experiments, that the erythrocyte interaction is species specific.

V. Transmission of the Erythrocyte Interaction. From the experiments described we have evidence of an intercellular interaction at a membrane separation of 4 μm. This interaction depends on the membrane being intact, and on its having a normal membrane potential and a supply of energy. These requirements are those of the Fröhlich theory. There is also evidence which can attribute specificity to the interaction. An additional finding which does not appear in the Fröhlich theory in its early form is that extended macromolecules have to be present in the suspending medium for an interaction to occur. An interaction does not occur when the cells are suspended in isotonic salt solutions. Nor does it occur significantly in serum (plasma with the fibrinogen removed by the clotting process) nor in solutions of human albumin (the major protein constituent in serum from normal subjects). Pohl (1980,1981) on the other hand has detected an attraction of inorganic particles, which he attributes to a Fröhlich process, in cells other than erythrocytes, without macromolecules being present in the suspending phase. But these experiments had to be done in media with a very low salt content and hence very low electrical conductivity. So there is no incompatibility with our findings. They suggest that Pohl is detecting a direct electromagnetic attraction of membranes for artificial particles of high dielectric constant; in our work the normal electrical conductivity of blood would tend to neutralize a direct electromagnetic coupling, unmasking a more subtle interaction.

Our initial observations suggested that the erythrocyte interaction was transmitted by fibrinogen, an extended macromolecule and not by albumin, a compact globular macromolecule. On the other hand it could have been a matter of molecular size, fibrinogen having a much higher molecular weight than albumin. This possibility is unlikely because we find a highly significant interaction with isotonic saline solutions of both polyvinylpyrrolidone, MW 360,000 daltons and Dextran 70, average MW 70,000 daltons (Sewchand, Roberts & Rowlands 1982) and with poly(ethylene oxide) MW 300,000 and 100,000 daltons. Solutions of fibrinogen in isotonic saline at physiological concentrations (3-4 gl^{-1}) give the largest deviation from Brownian movement which we have measured, whereas albumin even at concentrations of 80 gl^{-1} (a little above the physiological range) gave no interaction. This suggests that the secondary or tertiary structure of the macromolecules determines the ability to transmit the interaction. It may be even more subtle still, for when we manipulate fibrinogen, without altering the mass concentration, the large interaction may diminish or disappear completely (Boyd, Masri, Sewchand, Roberts & Rowlands: unpublished). The biochemistry of fibrinogen is, however, complex and controversial (Doolittle 1981; Plow & Edgington 1982) and so this finding needs cautious interpretation, but it does suggest high specificity of the type envisaged by Fröhlich (1972,1975) for interactions between macromolecular systems.

VI. Contractils. This is the name we use for the contractile fibrils described in the Introduction, which hold together a rouleau subjected to traction (Fig.2). Since, on release of the traction, the fibrils contract down to zero length and the pre-existing rouleau reforms, it suggested that the fibrils were more than viscoelastic; that the contraction was active and dependent on a source of energy from the cells. We therefore tested the hypothesis that this too was a Fröhlich interaction, by the methods employed in investigating deviations from Brownian movement (see above).

Traction is applied to the end cells of a rouleau by sucking them into two diametrically opposed micropipettes of internal diameter 3 μm. Static results are recorded by still photomicrography and the dynamical aspects by a television camera and recorder or by high speed cinémicrography (Rowlands, Eisenberg & Sewchand 1983). Contractils occur when cells are in their own plasma and in the two rouleau-inducing solutions, polyvinylpyrrolidone, MW 360,000 and Dextran 70, average MW 70,000. Other solutions have not yet been studied.

The properties of the fibrils vary significantly under the experimental manipulations used to investigate the deviations from Brownian movement (see above: Rowlands *et al.* 1982b). Glutaraldehyde-fixed cells do not form rouleaux but they sometimes stick together. There is no sign of a contractil as they are pulled apart. When the membrane potential is abolished using ionophore A23187 normal-looking rouleaux form but, on attempting to apply traction, they fall apart. It is different when the membrane potential is abolished by reduction of pH. Short fibrils can be produced on traction but they break quickly. Similar short, weak fibrils arise with metabolically depleted cells. Restoring pH, or repleting metabolically deprived cells, restores contractile fibrils but only partially. That the revival was not complete is to be expected. Interruptions of energy supply causes membrane changes (Lutz 1978). Moreover since electrical potential is energy, the membrane potential may be the final source of energy to the coherent excitations and if it is not fully restored, this too could prevent full recovery of the interaction. Glansdorff & Prigogine (1971) find, theoretically, that energy is a fundamental thermodynamic necessity for the stability of an organized structure such as a membrane, for its function and, of course, for its assembly.

It is not surprising that short, weak fibrils can form when some inactive erythrocytes are pulled apart. Macromolecules are the adhesive and these should form a thread as the cells separate, as would two toy balloons, coated with heavy oil, and then pulled apart. Further points can be made in favour of an active (energy-requiring) process. The fibrils from living cells can be very long (>100 μm) and they are very stable. Figs.4&5 are scanning electron micrographs: they were produced from contractils which had existed for at least half an hour (Rowlands *et al.* 1983). When a contractil breaks, the remaining cells are pulled all the way back into a rouleau. If the threads were just viscoelastic, they would shorten but not down to zero length. In Fig.2 the pipettes are slightly out of alignment. It will be noted that the outside contractils attached to the cells held in the pipettes are directed radially, towards the centre of the spherical parts of the pipette-held cells. As the pipettes are manipulated in and out of alignment the points of attachment move freely over the spheres without hysteresis and the direction of the contractil is always radial to

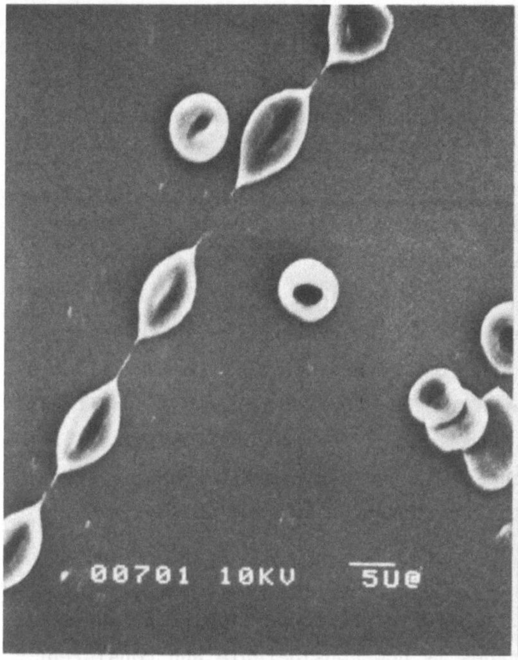

Fig.4.(left) Scanning electron microscope picture of a rouleau extended as in Fig.2. The macromolecular solution was slowly replaced with isotonic glutaraldehyde while tension was maintained on the pipettes. After being fixed by the glutaraldehyde the specimen was washed in distilled water, dried overnight and then gold-shadowed. (Line lower right is 5 μm long.)

Fig.5.(right) As in Fig.4 but at higher magnification. The break in the middle is artefactual but the discontinuities where the contractils meet the drawn-out cell membrane are always present. (Line lower right is 0.5 μm long.)

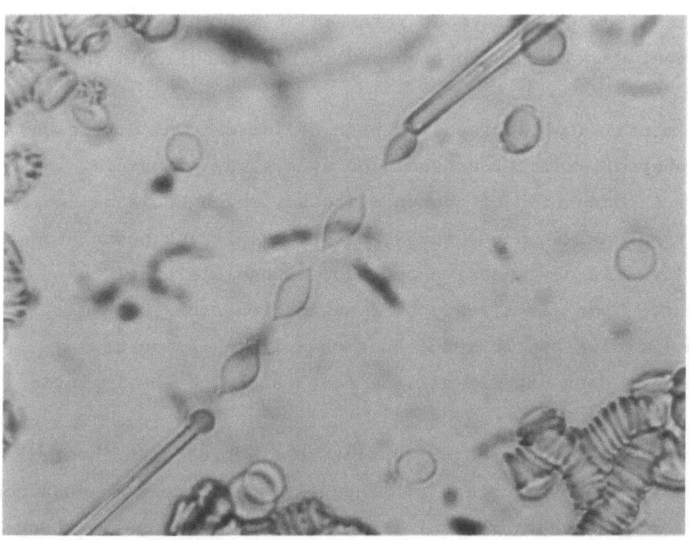

Fig.6. As in Fig.2. The negative pressure in the upper pipette has been reduced in magnitude. The point on the cell held in the upper pipette moves freely, with no lag, over the cell surface as the pipettes are rapidly moved out of and into alignment

the sphere. If the negative pressure in a pipette is reduced a little so that the external sphere sags (Fig.6) then a point arises at the attachment of the contractil; and this point also moves freely with the contractil as the alignment of the pipettes is changed. A passive viscoelastic thread would always attach at the point on the membrane from which it originated.

Rouleaux form in mixtures of cells from different species of mammals (Sewchand & Canham 1976). They are not random in their grouping of cells from the different species and this observation has been offered as evidence for specificity in the interaction of erythrocytes (Rowlands *et al.* 1982a,b; Sewchand & Rowlands 1983). We manipulated mixed rouleaux derived from human and rabbit erythrocytes which can be distinguished by their different sizes. On traction such a mixed rouleau will demonstrate weak contractils but it always separates at the junction between a rabbit cell and a human cell. But if cells are removed from a mixed rouleau until only cells from one species remain in the shorter rouleau, then, when pulled apart, normal contractils are seen. This clearly demonstrates a reduced adhesiveness between cells from different species.

VII. Discussion of the Various Experiments. The evidence supporting Fröhlich's theory of coherent excitations in cell membranes is now considerable and consistent (Fröhlich 1980; Webb 1980; Pohl 1980,1981 and the work described above). Biomedical scientists seem, on the whole, to be either ignorant of the theory or unwilling to consider it rationally. Perhaps Fröhlich will have to console himself with the sentiments of Max Planck: "An important scientific innovation rarely makes its way by gradually winning over and converting its opponents: it rarely happens that Saul becomes Paul. What does happen is that its opponents gradually die out and that the growing generation is familiarised with the idea from the beginning" (Bartlett 1968). Regrettably the opposition comes from the anonymous referees of scientific granting agencies which, as a consequence, deny the support needed to test Fröhlich's theory and its predictions.

Admittedly not one experimental approach has yet been decisive. One direct way of proving the existence of coherent excitations would be by a study of muscle. As it contracts there should be a condensation of phonon momentum, which would be detectable by slow neutron spectroscopy, along, perhaps, with the solitons postulated by Davydov (1977,1979). But it would be expensive and, so, impossible until the granting agencies see the light. There is already some suggestion from the Raman spectroscopy of active muscle (Pézolet *et al.* 1980) that the sliding filament hypothesis of muscular contraction does not fit all the facts and might be improved in the light of the Fröhlich theory.

The opposition is mainly of the "it cannot be" variety but a proffered chemical explanation of the findings is worth considering. The suggestion is based on enhanced water structuring near a surface and the formation of surface-associated gels. Long chain molecules might attach to the cell surface and successive associations with other molecules could form a shell of loosely-attached gel structure near the surface which, in turn, could interact with the gel around another cell. Such gels are affected by pH and ionic strength. It is stated that this interacting network could contract at the expense of chemical energy, though a source of chemical energy is not suggested. There is, of course, ample evidence of "structured" water in cells (Clegg 1981) and the adhesion of rouleaux has for twenty years been explained by the attachment of long-chain molecules to receptor sites on the membranes of the erythrocytes (Katchalsky *et al.* 1959; Rowlands & Skibo 1972). To fit the experimental findings this gel structure would have to extend 2 μm outside the membrane, so that when two membranes came within 4 μm of each other the gels could intermingle. But by what mechanism is *chemical* energy supplied to the gel at this moment to cause its contraction? Gels and biological macromolecules are sensitive to changes in electrolyte concentration (Cadman *et al.* 1982) but in the experiments in which the membrane potential was abolished the pH changes were modest and the findings were confirmed, without pH change, by treating the membranes with an ionophore to render them leaky. The presence of such a thick layer of gel on the surface of living cells and its absence on dead cells, microspheres, cells with no membrane potential and cells with no energy would have shown up in the measurements of the drag coefficients (Rowlands *et al.* 1982b, p.55). The standard error of the mean value of all the measurements under these various conditions is less than 4%. Further analysis of the individual figures reveals no pattern which supports the presence of such a gel. Other objections could be raised from many observations on the physical, chemical and biological properties of blood but our own, unpublished, experiments on small lymphocytes are worth quoting. Under identical conditions to those in the experiments on erythrocytes, the distribution of small lymphocytes is completely random as measured by the "nearest neighbour" technique (Clark & Evans 1954). An occasional small lymphocyte will become amoeboid, another will react strongly with platelets but the majority execute Brownian movement. Two lymphocytes may approach each other by chance. They rarely touch and if they do they quickly separate, presumably driven apart by their electrical surface charge. Yet a passive chemical gel mechanism should apply to lymphocytes as to erythrocytes.

The necessity of extended macromolecules to transmit the erythrocyte interaction is intriguing. Poly(ethylene oxide) is the most extended macromolecule (Stone & Stratta 1967) with which we have found an erythrocyte interaction (Rowlands & Long: unpublished). The minimum concentration of poly(ethylene oxide) MW 300,000 with which we detect an erythrocyte interaction is 0.005 gl^{-1}. A rough calculation shows that at this concentration the volume of each molecule is associated with 7000 volumes of water and that the mean separation between molecules is 34 nm. This separation is too great for gel formation. We have speculated that the erythrocyte interaction is mediated by quantum exchange of quasi-particles (Rowlands *et al.* 1982b). Davydov (1979) has offered theoretical evidence for the transmission of the energy of the hydrolysis of ATP from point to point, without loss, in cell structures by solitons moving in the α-helices of proteins. Solitons are quasi-particles which occur in non-linear systems and which, because of topological constraints, do not dissipate their energy (Bullough & Dodd 1977; Lamb 1980). Membranes and macromolecules certainly involve nonlinearities and so may sustain soliton transmission.

Del Giudice, Doglia & Milani (1982a,b) have combined the theories of Fröhlich and of Davydov into a model in which there is a "tight interplay between coherent electric vibrations and vibrational solitons". Of immediate interest is their proposal of attractive forces between macromolecules in a Fröhlich field which could lead to their alignment, aggregation and even polymerization.

If such filamentation is involved in the erythrocyte interaction at a distance it is likely also to be the explanation for contractils. Although a "chemical" gel explanation is much more reasonable in this case, it can be refuted, as above, in all respects except the lack of propinquity of the molecules. But the recoil to zero length and the absence of a fixed point of attachment of the contractil to the membrane make a gel hypothesis unlikely. The properties of contractils, together with the deviation from Brownian movement, forms a substantial body of experimental evidence for Fröhlich's hypothesis. Scrutiny of Fig.5 shows a gap in the middle of the contractil. This is an artefact of the preparation, but at each end of the contractil, just as it widens out to merge with the distorted contours of the cell, there is a very small gap and this occurs in all the pictures (Rowlands *et al.* 1983). In the better pictures a small thinner thread can be seen bridging the gap. There has to be something there because what is seen in the picture is gold which has been sputtered onto the cells and onto the contractils after they have been fixed in glutaraldehyde, washed in water and then dried. There is certainly some sort of singularity where membrane and contractil come in opposition and perhaps it is a gap across which quasi-particles are tunnelling? On the original pictures there is also an indication of a cross-structure to the contractils.

A Fröhlich mechanism will explain many presently obscure processes in living organisms. Fröhlich himself has discussed enzyme activity (Fröhlich 1970,1975), the interaction of cells with electromagnetic fields (Fröhlich 1980) and the problem of cancer (Fröhlich 1977), as have others (e.g. Cooper 1981). Vectorial ordered movement such as that of receptosomes (Pastan & Willingham 1981) and the "automated-factory-like" processing of proteins from the endoplasmic reticulum by the Golgi apparatus (Rothman 1981) now have an explanation as does the whole process of protein synthesis from DNA transcription to protein assembly. That is to say there is a mechanism which offers specific tensor interactions at considerable distances on the scale of a living cell.

The multitude of ordered processes operating at any one time in a cell and the additional processes which can be turned on are more complicated than an industrial factory, more complicated perhaps than a whole industry and it is inconceivable that they are not controlled. There are well documented communications between cells (Loewenstein 1979) which are probably necessary but far from sufficient. The Fröhlich process could add sophistication to Loewenstein's interaction but it is still an inadequate explanation, as it stands, for the function of a whole organ or organism.

VIII. Speculation: A Second Nervous System? This brings me back to the beginning of this paper and the influence of electromagnetic fields on healing bone. An orthopaedic surgeon, R.O. Becker has done many studies from the regrowth of amputated limbs in salamanders to the regrowth of bone in man (Marino & Becker 1977; Becker 1979). He explains his work in terms of *Neurotrophism* which is defined as "interactions between nerves and other cells which initiate or control molecular modifications in the other cell" (Hubbard 1974). Interruption of nerve supply and various neurological disorders have effects additional to the well-known motor and sensory paralysis. There are changes in the skin and other tissues. In animals, section of the lingual nerve causes the taste buds to disappear and they reappear when the nerve is rejoined. Leprosy is a "neurotrophic" disease. Ranney (1973) describes an interesting case of sensory loss in all fingers of one hand except the middle finger. The other fingers had shortened to the length of the first finger joint but with remnants of the fingernails still present. Ranney has also shown the value of neurovascular transplants in halting such progressive "antigrowth" (Ranney & Lennox 1978). Head injuries sometimes cause bone to be deposited "all over the place" (C. Frank, private communication) and there is a well-documented condition, melorheostosis Léri (Léri & Joanny 1922), in which bone is irregularly thickened giving the appearance on X-ray of candle-wax flowing down the margin of the bone. It is associated with an embryonic developmental error in the nervous system but is not related in its distribution with any major nerves. Rather, it follows segmental sensory nerve supply of bone itself (Murray & McCredie 1979). These types of observations and his own led Becker to postulate a "second nervous system, electronic in nature, more primitive and basic to the well-known action potential" (Becker 1979). I would agree except with it being primitive and basic. Quite the reverse in my view!

In a scholarly article Fanchon Fröhlich (1977) has analysed the complexity of the language which must be involved in cellular information processing and communication. This analysis makes it quite clear that a DC system would be inadequate by many orders of magnitude for the neutrophic communication between neurons and somatic cells. But supposing Herbert Fröhlich is correct and cells *are* communicating at very high frequencies. Neurons, too, are cells and they may have adapted these frequencies to information processing. A communications engineer would be in paradise if he had a communication band extending say from 10^{10} to 10^{11} Hz. He could pack over a million frequency-modulated radio stations into this range or 150,000 television broadcasts. If we assume a relaxation time for cell mechanisms, other than those mediated by action potentials, to be of the order of one second, then there is ample time to transmit complex information; for 10^{-11} second is to one second as one second is to 3000 years. The action potential may be just a crude, fast transmitter of urgent messages and the "second nervous system" the important one, following the precept that anything important is not urgent.

The vibrations might be transmitted along the membranes of the nerve fibres (axons) but they would be interrupted by action potentials. It is more likely that microtubules in the axon are used. These are ordered helices of a protein, tubulin, which extend along the length of the axoplasm. Their function in nerve is unknown, though they are believed to be associated with axonal molecular transport, and in other cells microtubules are ascribed the function of a cytoskeleton. Del Giudice *et al.* 1982a,b) have explained Webb's results (1980) by mechanisms analogous with well-established effects in nonlinear optics (NLO). Self-focussing and filamentation are other NLO effects. Del Giudice *et al.* suggest that Fröhlich vibrations may, similarly, be focussed into microtubules (and even assemble them), thereby being transmitted over long distances, without absorption in the ever-present water.

In the central nervous system the neuroglial cells might be the centre of the second nervous system. Neuroglial cells are complex in structure and they outnumber neurons in the central nervous system by at least ten to one. Their function is obscure and hypotheses range from that of a mechanical support structure to the seat of memory (Galambos 1971). The membranes of glial cells invest those of neurons very extensively and they come closer together (10 nm) than any cells which are not actually in contact (Kuffler & Nicholls 1976; Orkand 1982; Picker & Goldring 1982). This gap is close enough for the tunnelling of quantum quasi-particles between the two moieties of specialized cells of the central nervous system.

There is therefore an adequate structure for a "second nervous system" and the mechanism is there in the Fröhlich hypothesis of coherent excitations. I have developed this heretical speculation at some length because it may be easier to test experimentally than, say, coherent excitations as an explanation for the tensor interactions of the Golgi apparatus. Adey and his colleagues (Adey 1981; Lawrence & Adey 1982) have techniques for studying the interaction of high frequency electromagnetic fields with the central nervous system. Developments in phonon optics and the propagation of phonons in semi-conductors at the Bell Laboratories (Narayanamurti 1981) make it feasible to detect or to inject phonon excitations directly. The advantage of neurons over other cells is that their processes can be large (up to 1 mm in diameter in the giant axon of the squid and metres long in mammals) and many nerve fibres run together in large parallel bundles.

Even if these speculations turn out to be false, the Fröhlich hypothesis is unlikely to share the same fate. The evidence in its favour grows more compelling. As evidence accumulates, the theory will develop as did quantum mechanics after Bohr's theory of the hydrogen atom. Fröhlich's theory will be to quantum biology in the 21st century as the Bohr theory is to quantum physics and chemistry in the 20th.

Acknowledgment: The work was supported by the Medical Research Council of Canada. I am grateful to many people for patiently listening to these ideas, but in particular I must thank my close collaborator Dr. Lionel S. Sewchand. Had it not been for him I would have abandoned the work at an early stage!

References

Adey, W.R. (1981) Tissue interactions with nonionizing electromagnetic fields. Physiol. Rev. *61*, 435-514.

Bartlett, J. (1968) Familiar Quotations. 14th Edition, Little-Brown, Boston.

Becker, R.O. (1979) The significance of electrically stimulated osteogenesis: more questions than answers. Clin. Orthop. *141*, 266-274.

Bullough, R.K. and Dodd, R.K. (1977) Solitons; *in* Synergetics. Edited by H. Haken, Springer-Verlag, Berlin.

Cadman, A.D., Fleming, R. and Guy, R.H. (1982) Diffusion of lysozyme chloride in water and aqueous potassium chloride solutions. Biophys. J. *37*, 569-574.

Canham, P.B., Jay, A.W.L. and Tilsworth, E. (1971) The rate of sedimentation of individual human red blood cells. J. Cell Physiol. *78*, 319-332.

Clark, P.J. and Evans, F.C. (1954) Distance to nearest neighbour as a measure of spacial relationships in populations. Ecology *35*, 445-453.

Clegg, J.S. (1981) Intracellular water, metabolism and cellular architecture. Collective Phen. *3*, 289-311.

Cooper, M.S. (1981) Coherent polarisation waves in cell division and cancer. Collective Phen. *3*, 273-287.

Davydov, A.S. (1977) Solitons as energy carriers in biological systems. Studia Biophys. *62*, 1-8.

Davydov, A.S. (1979) Solitons in molecular systems. Physica Scripta *20*, 387-394.

deHaas, W.G., Lazarovici, M.A. and Morrison, D.M. (1979) The effect of low frequency magnetic fields on the healing of the osteotomized rabbit radius. Clin. Orthop. *145*, 245-251.

deHaas, W.G., Morrison, D.M. and Watson, J. (1980) Non-invasive treatment of ununited fractures of the tibia using electrical stimulation. J. Bone Joint Surg. *62B*, 465-470.

Del Giudice, E., Doglia, S. and Milani, M. (1982a) Fröhlich waves, self-focussing and cytoskeletal dynamics. Phys. Lett. *90A*, 104-106.

Del Giudice, E., Doglia, S. and Milani, M. (1982b) A collective dynamics in metabolically active cells. Physica Scripta *26*, 232-238.

Doolittle, R.F. (1981) Fibrinogen and fibrin. Scient. American *245*, 126-135.

Einstein, A. (1926) Investigations on the theory of the Brownian movement. Edited by R. Fürth (Translated by A.D. Cowper) Dover Publications, New York, N.Y.

Fåhraeus, R. (1929) The suspension stability of blood. Physiol. Rev. *9*, 241-274.

Fröhlich, F. (1977) The linguistic structure and the chromosome genetic code and language. *in* Synergetics. Edited by H. Haken, Springer-Verlag, Berlin.

Fröhlich, H. (1968) Long range coherence and energy storage in biological systems. Int. J. Quantum Chem. *2*, 641-649.

Fröhlich, H. (1970) Long range coherence and the action of enzymes. Nature *228*, 1093.

Fröhlich, H. (1972) Selective long range dispersion forces between large systems. Phys. Lett. *39A*, 153-154.

Fröhlich, H. (1975) The extraordinary dielectric properties of biological materials and the action of enzymes. Proc. Nat. Acad. Sci. USA *72*, 4211-4215.

Fröhlich, H. (1977) Biological control through long range coherence. *in* Synergetics. Edited by H. Haken, Springer-Verlag, Berlin.

Fröhlich, H. (1980) The biological effects of microwaves and related questions. Adv. Electron. Electron Phys. *53*, 85-152.

Galambos, R. (1971) The glia-neuronal interaction: some observations. J. Psychiat. Res. *8*, 219-224.

Glansdorff, P. and Prigogine, I. (1971) Thermodynamic theory of structure, stability and fluctuations. John Wiley and Sons, Ltd., London.

Hubbard, J.I. (1974) Trophic Functions. *in* The Peripheral Nervous System. Edited by L. Guth, Plenum Press, New York, N.Y.

Katchalsky, A., Danon, D., Nevo, A. and deVries, A. (1959) Interactions of basic polyelectrolytes with the red blood cell. II. Agglutination of red blood cells by polymeric bases. Biochim. Biophys. Acta *33*, 120-138.

Kernick, D., Jay, A.W.L., Rowlands, S. and Skibo, L. (1973) Experiments on rouleau formation. Can. J. Physiol. Pharm. *51*, 690-699.

Kuffler, S.W. and Nicholls, J.G. (1976) From neuron to brain. Sinauer Associates, Sunderland, Mass.

Lamb, G.L. Jnr. (1980) Elements of soliton theory. John Wiley, New York, N.Y.

Lawrence, A.F. and Adey, W.R. (1982) Nonlinear mechanisms in tissue-electromagnetic field interactions. Neurol. Res. *4*, 115-153.

Léri, A. and Joanny, J. (1922) Une affection non décrite des os hyperosteoses en coulée sur toute la longueur d'un membre ou melorhéostose. Bull. Soc. Méd. Hôp. Paris *46*, 1141-1145.

Loewenstein, W.R. (1979) Junctional intercellular communication and the control of growth. Biochim. Biophys. Acta Cancer Rev. *560*, 1-65.

Lutz, H.U. (1978) Vesicles isolated from ATP-depleted erythrocytes and out of thrombocyte-rich plasma. J. Supramol. Struct. *8*, 375-389.

Marino, A.A. and Becker, R.O. (1977) Biological effects of extremely low frequency electric and magnetic fields. Physiol. Chem. Phys. *9*, 131-147.

Murray, R.O. and McCredie, J. (1979) Melorheostosis and the sclerotomes: a radiological correlation. Skeletal Radiol. *4*, 57-71.

Narayanamurti, V. (1981) Phonon optics and propagation in semi-conductors. Science *213*, 717-723.

Orkand, R.K. (1982) Signalling between neuronal and glial cells. *in* Neuronal-glial interrelationships. ed. T.A. Sears, Springer-Verlag, Berlin.

Overbeek, J.T.G. (1952) *in* Colloid Science, Vol. I. Edited by H.R. Kruyt, Elsevier Scientific Publishing Co., Amsterdam.

Pastan, I.H. and Willingham, M.C. (1981) Journey to the centre of the cell: role of the receptosome. Science *214*, 504-509.

Pézolet, M., Pigeon-Gosselin, M., Nadeau, J. and Caillé, J.-P. (1980) Laser Raman Scattering: A molecular probe of the contractile state of intact single muscle fibres. Biophys. J. *31*, 1-8.

Picker, S. and Goldring, S. (1982) Electrophysiological properties of human glia. Trends in NeuroSciences *5*, 73-76.

Plow, E.G. and Edgington, T.S. (1982) Surface markers of fibrinogen and its physiologic derivatives revealed by antibody probes. Sem. Thromb. Hemost. *8*, 36-56.

Pohl, H.A. (1980) Oscillating fields about growing cells. Int. J. Quantum Chem. *7*, 411-431.

Pohl, H.A. (1981) Natural electrical RF oscillation from cells. J. Bioenerg. Biomembr. *13*, 149-169.

Ranney, D.A. (1973) The hand in leprosy. The Hand *5*, 1-8.

Ranney, D.A. and Lennox, W.M. (1978) The protective value of neurovascular island pedicle transfer in hands partially anaesthetic due to ulnar deviation in leprosy. J. Bone Joint Surg. *60A*, 328-334.

Rothman, J.E. (1981) The Golgi apparatus: two organelles in tandem. Science *213*, 1212-1219.

Rowlands, S., Eisenberg, C.P. and Sewchand, L.S. (1983) Contractils: quantum mechanical fibrils. J. Biol. Phys. (in press).

Rowlands, S., Sewchand, L.S. and Enns, E.G. (1982a) Further evidence for a Fröhlich interaction of erythrocytes. Phys. Lett. A, *87*, 256-260.

Rowlands, S., Sewchand, L.S. and Enns, E.G. (1982b) A quantum mechanical interaction of human erythrocytes. Can. J. Physiol. Pharm. *60*, 52-59.

Rowlands, S., Sewchand, L.S., Lovlin, R.E., Beck, J.S. and Enns, E.G. (1981) A Fröhlich interaction of human erythrocytes. Phys. Lett. A, *82*, 436-438.

Rowlands, S. and Skibo, L. (1972) The morphology of red-cell aggregates. Thromb. Res. *1*, 47-58.

Sewchand, L.S. and Canham, P.B. (1976) Induced rouleaux formation in interspecies populations of red cells. Can. J. Physiol. Pharm. *54*, 437-442.

Sewchand, L.S., Roberts, D. and Rowlands, S. (1982) Transmission of the quantum interaction of erythrocytes. Cell Biophysics *4*, 253-258.

Sewchand, L.S. and Rowlands, S. (1983) Specificity of the Fröhlich interaction of erythrocytes. Phys. Lett. A, (in press).

Sewchand, L.S., Rowlands, S. and Lovlin, R.E. (1982) Resistance to the Brownian movement of red blood cells on flat horizontal surfaces. Cell Biophysics *4*, 41-46.

Smoluchowski, M. v. (1917) Investigation into a mathematical theory of the kinetics of coagulation of colloidal solutions. Zeit. Phys. Chim. *92*, 129-168.

Steinberg, M.S. and Wiseman, L.L. (1972) Do morphogenetic tissue rearrangements require active cell movements? J. Cell Biol. *55*, 606-615.

Stone, F.W. and Stratta, J.J. (1967) Encyclopedia of Polymer Science and Technology *6*, 103-145.

Webb, S.J. (1980) Laser-Raman spectroscopy of living cells. Phys. Rep. *60*, 201-224.

Intracellular Water, Metabolism and Cell Architecture: Part 2

JAMES S. CLEGG
Laboratory for Quantitative Biology, University of Miami, Coral Gables, Florida 33133, USA

INTRODUCTION

Biologists seem to pay little attention to physical theories that attempt to explain living systems without detailed reference to the elaborate structure that we know exists, notably at the cellular and subcellular levels. I suspect this explains, at least to some extent, why Fröhlich's theory of coherent excitations (See Fröhlich, 1980) has almost gone unnoticed by the biological community. Therefore, the present chapter will attempt to describe what we know about certain features of the structure and function of eucaryotic cells, selecting those aspects that seem to be most relevant to the theory.

The title carries "part 2" because an initial attempt in that direction was made three years ago (Clegg, 1981a). At that time it was pointed out that an intricate network of protein structures known as the "cytoskeleton", and notably the then very recently discovered microtrabecular lattice, extends from the cell surface throughout the cytoplasm. Evidence was presented that suggested that this architecture might be intimately associated with the metabolic machinery of cells and that it also determined the physical properties of a large fraction of intracellular water by virtue of its vast surface area. I proposed that all these relationships could play a significant role in the establishment and propagation of the coherence predicted by Fröhlich's theory. In the following section I will reconsider the essential features of the ideas presented in part 1 (Clegg, 1981a) and evaluate them in the light of research performed since that time. It should be stressed that space limitations prevent anything approaching a review of these very active research areas. Only recent reviews and key articles that provide access to the literature can be cited. As pointed out in the previous attempt (Clegg, 1981a) I claim no expertise in phy-

Coherent Excitations in Biological Systems
Ed. by H. Fröhlich and F. Kremer

sics in general, or in coherent phenomena in particular, and am in no position to support or refute the theory under consideration here. My objective is to provide a cellular point of view.

CELLULAR ARCHITECTURE

In the present context the group of cellular structures that appear to merit closest attention are the cytoskeletal components: microtubules (about 250Å diameter) made from the protein tubulin; microfilaments (about 60Å diameter) made from the protein actin; intermediate filaments (about 100Å diameter) whose protein composition varies. These and related structures have been the object of an enormous amount of research, much of which has recently been considered in the two published volumes of a Cold Spring Harbor Symposium (Albrecht-Buehler, 1982). In addition to these, the work of Porter and associates has indicated the existence of a highly branched three dimensional network that not only extends throughout the cytoplasm but is believed to surround virtually all cytoplasmic structures with the possible exception of mitochondria (see Porter and McNiven, 1982). This structure, abbreviated "MTL", contains a large amount of actin which appears to be its major structural protein; however, a wide variety of different proteins are also closely associated with it (see Schliwa, et al. 1981) a point to which I will return. The relationship between actin microfilaments and the MTL is controversial and I should point out that alternative views of the actin networks in cells have been presented (see Heuser & Kirschner, 1980; Albrecht-Buehler, 1982). My position is that Porter and colleagues have made a very strong case for the reality of the MTL, and I proceed here on that basis.

Of great importance to this discussion are extensive connections made between plasma membrane and the actin networks (see Poste and Nicolson, 1981; Avnur & Geiger, 1981; Penman et al. 1982). It is well documented that many extracellular signals, a number of hormones for example, exert their influence by first binding to receptors at the cell surface. This binding is followed by cytoskeleton - mediated events and/or the production of "second messengers" (cAMP, cGMP) via activation of the appropriate cyclase enzyme, also in or near the cell membrane. Such "transmembrane modulation" plays a dominant role in cellular function. Because the MTL makes direct or indirect connections with these cell surface receptors as well as nearly all other cytoplasmic structures this system provides an excellent means of intracellular communication.

But the situation is made even more interesting by the detailed work of Penman and associates (see Penman et al. 1982 for review). They demonstrated that the functional units of protein synthesis, polysomes, are also contained within or on the MTL (those polysomes that are not attached to endoplasmic reticulum). Consequently, these key cell components are structurally connected with cell surface receptors, as well as other cell structures, via the MTL. Such connections clearly must have important functional consequences. The Penman research group has also recently shown that connections exist between nuclear chromosomes, the nuclear cytoskeleton and the MTL (Capo et al. 1982). Those observations are of potentially enormous importance because they allow for the possibility of genetic control, and other nuclear events such as those involved with cell division, all potentially mediated via the actin networks.

The picture that emerges from this brief description (diagrammed in Fig. 1) is one in which essentially all of the ultrastructurally recognizable structures in cells are interconnected by the MTL. Dotted lines represent the MTL in association with a single cell surface receptor. However, it should be realized that a very large number of these receptors exist and, although direct evidence is lacking, it is possible that most of these are at least indirectly associated with the MTL. Relationships between enzymes and the MTL will be considered later in this paper.

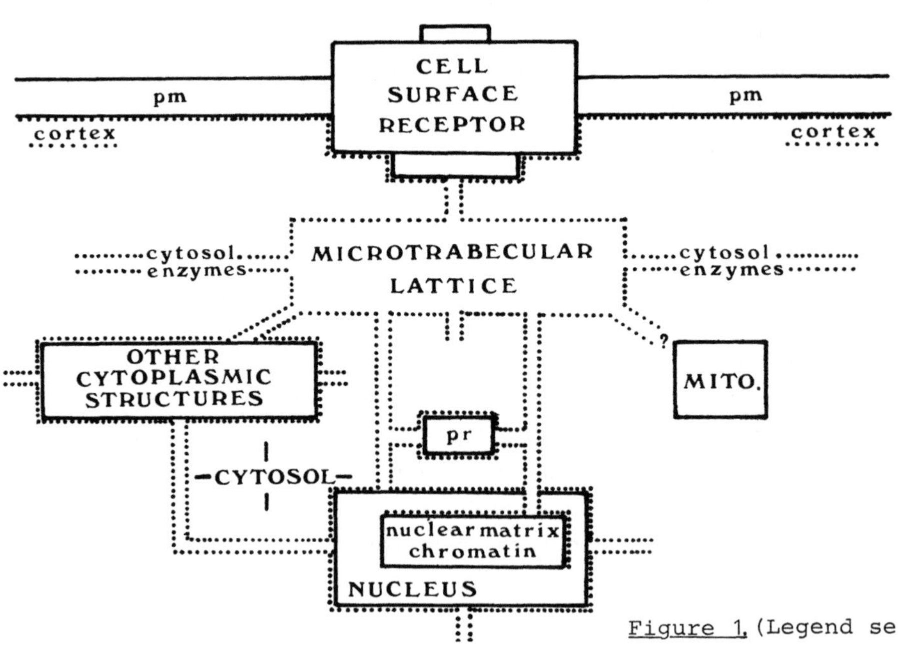

Figure 1. (Legend see p. 165)

An analogy, perhaps too fanciful, would describe the cell as an electronic schematic diagram in which the various electronic components (cell organelles, etc.) are interconnected and controlled by inputs via wiring (actin networks, MTL) from other parts of the instrument (cell) including major controlling switches (cell surface and intracellular receptors). Input from one small area (the cell surface, for example) can be amplified enormously and rapidly conveyed to the appropriate part of the cell.

Before leaving architectural matters it is appropriate to briefly consider the dynamics of this system. I concentrate here on the actin-based components. Many studies have been carried out on the "turnover" of actin structures in vitro (see Korn, 1982) and in vivo (see Kreis and Birchmeier, 1982). In the test tube single molecules of globular actin (G-actin) can, under suitable conditions, be polymerized into filaments (F-actin). The growth of such filaments takes place at the two ends, one being faster than the other. The net rate of growth in vitro can be as fast as 90 G-actin monomers added per second per filament, amounting to an increase in length of about 3000Å (0.3 μm) per second per filament. Recent work has clearly shown that the turnover of the actin networks in living cells is constant and also rapid, although precise numbers are apparently not available.

The energetics of polymerization is of some interest to us (see Korn, 1982). Each actin monomer (G-actin) contains 1 molecule of bound ATP, the free energy currency in living cells. It is a puzzling fact that this ATP does not provide the free energy for polymerization;however, it is split into ADP (which remains bound) and inorganic phosphate (which is released) after F-actin is formed. Thus, the formation of actin filaments involves considerable free energy input, well in excess of that required for polymerization. Although several proposals have been made, the role of this energy expenditure remains obscure. In the final section of this chapter I will speculate on how the energetics of actin polymerization might be related to Fröhlich's theory.

Figure 2 summarizes several features of the foregoing discussion and provides some additional aspects of cell architecture, admittedly oversimplified and not entirely accurate.

Figure 1. Diagrammatic Representation of the Microtrabecular Lattice and its relationship to other Cell Structures (pm = plasma membrane, pr = polysomes; "cytosol" refers to the aqueous compartments of the cytoplasm)

Figure 2A represents an actin filament associated with the plasma membrane. These associations are extensive and probably involve specific linkage protein (LP) such as vinculin (Avnur & Geiger, 1981; Isenberg et al. 1982) or several others (see Albrecht-Buehler, 1982) connected to an integral membrane protein (IP). It should be noted that a wide variety of specific surface receptors have been shown to be associated with actin networks, both structurally and functionally (for example, Hunt and Hood, 1982; Snyderman and Goetzel, 1981; Conti-Tronconi, 1982; Puca and Sica, 1981; Prives et al. 1982; Pribluda and Rotman, 1982). Thus, the integral protein (IP) in Fig. 2A can also be viewed as such.

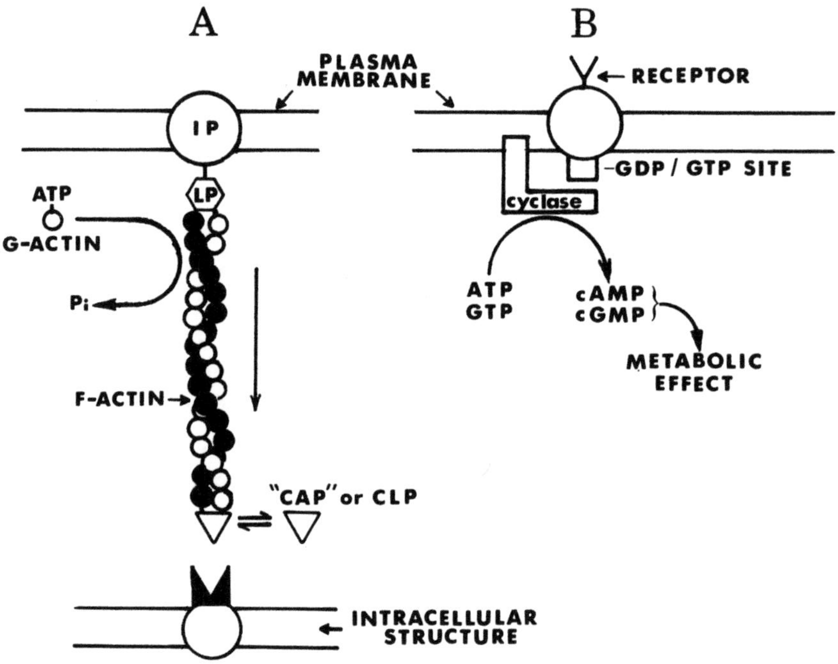

Figure 2. Diagram of Actin Filaments (A) and the Transmembrane Control of Cyclic Nucleotide Production (B)

There is some evidence that filament growth occurs by addition of G-actin to the membrane-associated end of the filament (see DeRosier and Tilney, 1982). Many proteins have been implicated in the control of filament growth and connection to other cell parts represented (not very precisely) as "cap" and cross-linking proteins (CLP). These affect filament length and the organization of three dimensional networks. The magnitude of this activity is considerable since actin can make up as much as 15-20% of total cell protein.

Figure 2B depicts the transmembrane modulation process discussed previously in this section. This complex mechanism is turned on or off depending upon the presence or absence of specific ligand bound to the receptor. The cyclase enzyme (adenyl or guanyl) when activated produce the cyclic nucleotides (second messengers). These systems also require significant energy input when active: for each mole of cyclic nucleotide produced 3 "high energy" phosphoryl groups are utilized, one of which involves GTP at a regulatory site (Fig. 2B). It is of some interest in the context of this discussion that the cytoskeletal components have very recently been implicated in the activities of these systems as well (Sahyoun et al. 1981; Cooper, 1982; Rasenick, 1981; Westermark and Porter, 1982; Browne et al. 1982). That feature is not shown in Fig. 2B. I will return to these important matters in the last section of this paper.

ENZYMES AND CELLULAR ARCHITECTURE

It has long been known that some enzymes in cells are integral parts of various membrane systems or other structures such as ribosomes and glycogen granules. However, most of intermediary metabolism has traditionally been considered to result from enzymes freely dissolved in solution within the aqueous compartments of cells, the "soluble enzymes". That conception, based largely on cell disruption studies, is currently being questioned and reexamined because of the increasing evidence that few enzymes exist free in solution when intact cells are examined by appropriate techniques. Access to the extensive literature on this and related matters is available through recent reviews (Kempner, 1980; Masters, 1981; Clegg, 1981a,b; Fulton,, 1982) and papers (Weber and Bernhard, 1982; Klyachko et al. 1982; Morton, et al. 1982; Clegg, 1982a; Mansell & Clegg, 1983). I believe that an examination of this evidence clearly indicates that the great majority of intracellular macromolecules, including enzymes are probably not free in solution, but are instead loosely associated with formed structure. The question of their location is of some significance, but remains partially unanswered. In skeletal muscle tissue there is very good evidence for the association of most glycolytic enzymes with the thin filaments, which are chiefly actin (see Masters, 1981; Morton et al. 1982). Moreover, a recent study by Schliwa et al. (1981) strongly suggests that many cytoplasmic proteins are loosely associated with the actin networks of non-muscle cells. That paper, which combined ultrastructural studies with an analysis of cellular proteins in normal and detergent-treated cells, in conjunction with evidence just cited, provides good reason to postulate the extensive organization of enzymes along (or within) actin networks of cells. It is quite likely, furthermore, that enzymes of a given metabolic pathway are organized into loosely associated multienzyme complexes. That is, these enzymes may be in close proximity to each other as well as being associated with cell architecture. The basis for that possibility comes from a number of studies on metabolic flux measurements, calculations of metabolite-transit times, and kinetic tracer studies of intracellular pools (see Kempner, 1980; Masters 1981; and references therein).

Figure 3 shows a diagrammatic representation of the traditional (Fig. 3B) and "organized" (Fig. 3C) conception of the disposition of enzymes in the aqueous compartments of cells.

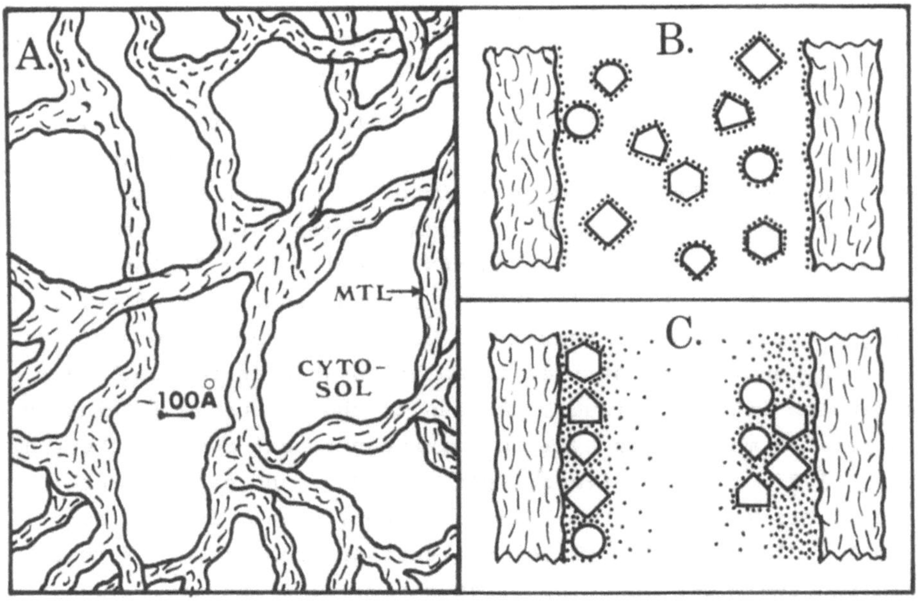

Figure 3. Diagrammatic Representation of the Microtrabecular Lattice (A) and its Possible Relationships to Cytosol Enzymes (B,C)

My own bias leads me to use the MTL (shown in Fig. 3A) as the major architectural site for such organization. Its variable thickness (50-200Å) is consistent with the description in Fig. 3C. Of course, locations other than actin networks could also be involved. The advantages of such organization, compared with a random solution-based metabolism, have frequently been pointed out (see Masters, 1981).

At this point, then, a reasonable case has been made for the existence of continuity of structure from cell surface throughout the cytoplasm and extending into the nucleus. If the view represented in Fig. 3C is correct the enzymes of intermediary metabolism, by virtue of their self-association (for a given pathway) and their close relationship to this architecture, represent a very high degree of organization. It seems possible that the entire metabolic machinery of eucaryotic cells is involved in this organization. In fact, there is reason to believe that even many small molecules (metabolites) may also be associated with macromolecules and/or cell structures, perhaps existing in "free and bound" forms. That is certainly true for the important coenzymes NAD^+ and $NADP^+$ and their reduced forms, and Wheatley & Inglis (1981) have presented and reviewed extensive evidence for amino acid binding in cells. Finally, Kempner (1980) has described experiments that show the nonrandom distribution of fluorescein (MW about 350) in the aqueous compartments of intact cells. The compartmentation of small metabolites as well as enzymes is a topic of much current interest This compartmentation has been described as "structural" (metabolites segregated by membranes) and "dynamic" (metabolites segregated by functional means). Current views are available in publications by Hess et al. (1980 and Sies (1982). It certainly appears that the molecular organization of cells is far more extensive than commonly believed.

The foregoing discussion leads us to a consideration of the aqueous compartments: those regions of cells in between the vast and complex architecture observable by microscopy.

INTRACELLULAR WATER AND THE AQUEOUS COMPARTMENTS

These compartments can be considered as those regions of eucaryotic cells which contain no demonstrable ultrastructure. Obviously, our understanding of these compartments has changed as more and more structure was demonstrated to exist. Thus, about 20 years ago the cytosol would inadvertently have included all cytoskeletal components as well as the MTL, simply because those structures were not known to exist. Therefore, our current view of the cytosol must consider the possible existence of organization which is not observable by current methods. Given this qualification, the cytosol would seem to be a relatively dilute aqueous solution containing only inorganic ions, small metabolites and, perhaps, a few macromolecules. That picture is much different from the traditional view (Fig. 3B). A hazy conceptual boundry exists between cytosol and the bordering non-aqueous ultrastructure because the system is dynamic, and the composition of the cytosol is likely to be constantly changing. A similar description can be given for the aqueous compartment of the nucleus, the "nucleoplasm.

Water dominates the composition of all these compartments of course, being, on a mole fraction basis, over 0.90 for all cell constituents. It is necessary to realize that this water is exposed to the truly vast surface area presented by cellular architecture. An example from the very recent work of Gershon et al. (1982) is illustrative. Using modern image analysis methods they estimated the surface area and volume of the MTL and associated cytoskeletal components in mammalian tissue culture cells. For a single cell of 16 μm diameter the surface area of these cytoplasmic structures is between 47,000 and 95,000 μm^2. This calculation excludes the nucleus which is taken to be 10 μm diameter. Thus, the total cytoplasmic volume is 75% of the total cell volume, or about 1620 μm^3. Therefore, the surface to volume ratio of the cytoplasm is about 30 to 60, and this does not consider the surface area of all of the other cytoplasmic architecture (membranes, granules, ribosomes etc.) which is very considerable (see Clegg & Drost-Hansen, 1978). The nucleus has its own "cytoskeletal" system which is also enormous (see Penman et al. 1982).

On this basis alone we should expect a large fraction of cell water to be in reasonably close proximity to ultrastructural surfaces (Fig. 3A). That is significant because water within 25-30Å of biological surfaces (see Lis et al. 1982) as well as non-biological ones (see Israelachvili, 1981) is known to exhibit physical properties that differ significantly from those of pure water. Such "vicinal water" Drost-Hansen 1978) is depicted by stippling in Fig. 3C.

Indeed, 25-30Å is a minimal distance over which the influence of the surface on water is "felt". Drost-Hansen (1978) has compiled evidence that this influence can extend over 500Å from such surfaces. If correct, then all cell water will be altered (see Fig. 3A). However, the traditional view of intracellular water has been that practically all of it has properties that are identical to those of pure water (reviewed by Clegg, 1982b). Indeed, much of current thought about cell structure and function is based on that description of intracellular water. In contrast, an increasing body of recent evidence strongly indicates that a large fraction of the water in living cells exhibits properties that are distinctly different from those of the pure liquid. Among the most compelling to me are the studies by Horowitz, Paine and associates (see Paine & Horowitz 1980; Horowitz and Pearson, 1981) who used an ingenious technique to show that the solvent properties of cytoplasmic water in amphibian oocytes differ profoundly from those of pure water. They argued that their results are probably valid for animal cells in general. It is very difficult to reconcile such results with the traditional view. These and other descriptions of cell water, including the traditional one, can be found in books edited by Keith (1979), Drost-Hansen & Clegg (1979) and Franks (1982).

I believe we are approaching at least a partial resolution of this long-standing controversy, and it is my view that the traditional position will undergo drastic modification. It is even possible that none of cell water has the properties of pure water, a view that has long been advocated by Ling (1980) and strongly supported by Negendank (1982) and Hazlewood and associates (see Hazlewood, 1979). All non-traditional views on cell water invoke the influence of intracellular surfaces. Thus, there is a sound basis for the perturbation of a large fraction of the water in living cells.

The significance of altered cellular water, compared with the pure liquid, can now be considered. If the view that virtually all metabolism occurs at or near the surfaces of cellular structures is accepted then it is evident that the enzymes that perform this role find themselves in a microenvironment that is clearly not a dilute aqueous solution. It is well known that the microenvironment can have a profound influence on the kinetic, thermodynamic and regulatory properties of enzymes (see Masters, 1981) and particularly the thorough and quantitative analysis by Welch and Keleti (1981) and Welch et al. (1982). Careri (1982) has given an impressive account of the importance of hydration to enzyme dynamics (also see Careri et al. 1979). These treatments leave little doubt that descriptions of cell metabolism and its regulation that are based on results from dilute solution studies in vitro are likely to be incomplete at best, and perhaps largely incorrect. The forces involved in enzyme-enzyme and enzyme-structure associations of the type postulated here (Fig. 3C) are not well understood (see Masters, 1981) although it is clear that they are not covalent. Because changes in water structure usually take place upon association of molecular sub-units into more complex aggregates, as in hydrophobic interactions for example, it is possible that water might participate similarly in the proposed enzyme-enzyme and enzyme-cell structure associations (Clegg, 1981a & 1982a,b). That possibility requires that the water proximal to both the ultrastructural surface and the enzyme has lower entropy (more "structure") than the water that is more distant from such surfaces. Association of the two might then be driven by the entropy increase attending the release of the more "structured water". Of course, other interactions are likely to be involved.

The description of cell structure and function given in the preceding sections of this chapter will now be considered in the context of Fröhlich's theory.

ON A CELLULAR BASIS FOR FRÖHLICH'S THEORY OF COHERENT EXCITATIONS IN BIOLOGICAL SYSTEMS

As I understand it, Fröhlich (1980) has proposed that strong electric fields ($\approx 10^7$ volt/m) in the cell membrane lead to strong electric polarization of membrane-bound molecules, notably proteins, which oscillate coherently leading to electric polarization waves. A certain minimum rate of "metabolic energy input" is required to establish that coherent behavior. To my knowledge the nature and location of energy input have not been clearly specified, although the cell membrane is obviously involved in this respect. From all that we know about cellular energetics, energy transactions almost always involve the removal of phosphoryl groups from ATP (or related ribonucleoside triphosphates). It would seem that the actin-networks in cells might be intimately related to the metabolic energy input required by the theory by virtue of their close proximity to cell membranes, their generally constant turnover, and the involvement of ATP in their formation (Fig. 2A). Recall that G-actin bound ATP is split after the actin monomer becomes part of the filament. Because this free energy transaction occurs after, and is not lost during polymerization this might provide a means of channeling the energy input required for coherence which arises due to "the sharing of energy by many macromolecules". Sufficient reason has been given in this paper to believe that the actin networks provide an excellent ultrastructural coupling device, a means of "storing energy" and propagating coherence throughout the cell.

Another ultrastructural feature of relevance is the ability of cells to control the rate of G-actin addition; thus, turnover is regulated and the system can be "switched" to higher or lower levels. Actin monomers appear to be added chiefly at nucleating sites near the cell surface (plasma membrane) to bring about filament growth, although there is not uniform agreement on that point. Since these sites often appear to be very near or directly connected to specific receptors in the plasma membrane (and/or transmembrane elements associated with them) very weak stimuli impinging on the cell surface could rapidly be transmitted throughout the cell (Fig. 2A,B). Such a system also allows for very large amplification so it is reasonable to speculate that this mechanism could be involved in the "non-thermal" microwave effects predicted by Fröhlich's theory. Such effects, although the subject of some debate, can either be positive or negative - that is, they stimulate or inhibit biological processes (or can have no effect). That is readily accounted for since the cell surface receptors could be activated or deactivated, depending on their status at the time of irradiation

(Fig. 2). Significant cellular effects could, therefore, result from very low levels of energy input. Moreover, if the proposed relationship between enzymes, the actin network and their connections is correct (Figs. 2,3c) then there is a direct link between such "sensitive domains" and the fundamental metabolic machinery from which so much of cellular structure and function originates. More on this later.

The involvement of cellular architecture has also been considered by Del Giudice et al. (1981, 1982a) but from a much more physical viewpoint, and within the context of Fröhlich's theory. Taken together with the more "biological" framework outlined here the foregoing speculation seems perhaps more reasonable. In addition, Berry (1981) has proposed an interesting electrochemical "model" of metabolism that shares several features in common with the Del Giudice group as well as my own views. Although Berry was apparently not aware of Fröhlich's theory, his model seems to be to be compatible with it.

Another feature of the theory under consideration concerns the possibility of frequency-selective long range interactions which arise from coherent excitation. Thus, an interesting possibility presents itself: the association of enzymes with the actin networks could involve their initial attraction by such attractive forces. Of some interest in this connection is the work of Starlinger (1967) who found that electrical stimulation of rat muscle in vivo increased the amount of the enzyme aldolase that was bound to the muscle structure. The effect "decayed" in less than 10 minutes post-stimulation. More recently, Clarke et al. (1980) have clearly shown that electrical stimulation increases the binding of most glycolytic enzymes to the actin filaments of bovine muscle. They offer several possible explanations, but point out that the mechanism of this electrically-induced enzyme binding is unknown. The Fröhlich effect could conceivably be involved. The theory also predicts that substrates will be attracted to their enzymes, provided that the former possess a vibrational frequency close to that of the coherent modes of the metastable enzyme. These possibilities will probably dismay the biochemist who is inclined to think only in terms of very short range interactions between enzymes, their regulators and substrates. As admitted earlier, I am in no position to critically judge Fröhlich's theory, but this single example indicates its potentially profound importance to metabolism and its control, should it be shown to be valid. One point seems clear: metabolic organization in intact cells is far far greater than previously believed.

A final consideration concerns the involvement of cellular water. I have presented a brief account of what I believe to be compelling evidence that a large fraction of intracellular water is sufficiently close to cellular architecture as to undergo changes in its physical properties compared with water in dilute aqueous solution. This "perturbed" water could play an important role in Fröhlich's theory. I previously suggested that the water within about 30Å from enzymes and ultrastructural surfaces might be involved in slowing the thermalization of energy released by the postulated coherently oscillating macromolecules and speculated on its possible participation in "non-thermal" microwave effects (Clegg, 1981a). Del Guidice et al (1982b), accepting the existence of the abundance of such interfacial water, have proposed that coherence among water dipoles should arise, depressing their usual thermal motion and forcing them to oscillate in phase. They go on to suggest that resonance between macromolecular dipoles ("excited by collective motion") and water dipoles surrounding the macromolecule may be involved in altering the properties of that surface water. It is not too unreasonable to propose that water in the vicinity of enzyme-actin network complexes must play a decisive role in metabolism and its regulation even though we only begin to dimly perceive the details of its involvement. Should Fröhlich's theory be proven correct it is clear that cellular water must be involved in a major way (Clegg, 1981).

CONCLUDING REMARKS

I am well aware of the fact that this chapter has contained a large amount of speculation. To some extent that speculation is magnified by space limitations which prevent thorough documentation of positions taken that are not in accord with current conventional wisdom, notably with regard to enzyme organization and the properties of cell water. On those two accounts I will stand by the views summarized here because I believe the evidence in their support is more than adequate. With regard to my interpretation of these views in terms of Fröhlich's theory I am much less confident. But it is fair to say that any attempt to deal with new and unusual ideas will most likely not result in a rigorous or even correct treatment. It appears that the idea of coherence in biological systems is quite new, at least to most biologists. As William James said: "When a thing was new people said - 'It is not true.' Later, when its truth became obvious, people said - 'Anyway, it is not important.' And when its importance could not be denied, people said - 'Anyway, it is not new." If we view these as three stages of scientific discovery it would appear to me that coherent excitations in living systems currently reside in the initial stage, or perhaps just beyond.

ACKNOWLEDGEMENTS

I am grateful to IBM Deutschland for travel and financial support during the meeting, and their warm and generous hospitality. Aided in part by research grant PCM 79-25609 from the U.S. National Science Foundation. The excellent clerical assistance of Mrs. Vivian Roe in the preparation of this paper is appreciated.

REFERENCES

Albrecht-Buehler, G. (ed.) 1982 "Organization of the Cytoplasm", Volume 46. 1046 pp. Cold Spring Harbor, N.Y.

Avnur, Z. and B. Geiger 1981 J. Mol. Biol. 153,361-379.

Berry, M.N. 1981 FEBS letters 134, 133-138.

Browne, C.L., A.H. Lockwood and A. Steiner 1982 Cell Biol. Intern. Rep. 6, 19-27.

Caper, D., K. Wan and S. Penman 1982 Cell 29, 847-854.

Careri, G. 1982 in "Biophysics of Water" (F. Franks, ed.) John Wiley and Sons, Ltd., Chichester (in Press).

Careri, G., P. Fasella and E. Gratton 1979 Ann. Rev. Biophys. Bioeng. 8, 69-97.

Clarke, F.M., F.D. Shaw and D.J. Morton 1980 Biochem. J. 186,205-109.

Clegg, J.S. 1981a Collect. Phenom. 3, 289-312.

Clegg, J.S. 1981b J. Exp. Zool. 215, 303-313.

Clegg, J.S. 1982a in "Cold Spring Harbor Symposia on Quantitative Biology" 46, 23-37, Cold Spring Harbor, N.Y.

Clegg, J.S. 1982b in "Biophysics of Water" (F. Franks, ed.) John Wiley and Sons, Ltd., Chichester (in Press).

Clegg, J.S. and W. Drost-Hansen 1978 in: "The Physical Basis of Electromagnetic Interactions with Biological Systems" (L.S. Taylor and A.Y. Cheung, eds.) pp. 121-131. HEW Publication (FDA) 78-8055. Washington.

Cooper, D.M.F. 1982 FEBS letters 138, 157-163.

Conti-Tronconi, B.M., M.W. Hunkapiller, and M.A. Raftery 1982 Science 218, 1227-1229.

Del Giudice, E., S. Doglia and M. Milani 1981 Phys. Lett. 85A, 402-404.

Del Giudice, E., S. Doglia and M. Milani 1982a Phys. Lett. 90A, 104-259.

Del Giudice, E., S. Doglia and M. Milani 1982b Physica Scripta 26; 232-238.

DeRosier, D.J. and L.G. Tilney 1982 in: "Cold Spring Harbor Symposium on Quantitative Biology" 46, 525-540. Cold Spring Harbor, N.Y.

Drost-Hansen, W. 1978 Phys. Chem. Liq. 7, 243-348.

Drost-Hansen, W. and J.S. Clegg (eds.) 1979 "Cell-Associated Water" 440 pp. Academic Press, N.Y.

Franks, F. (ed.) 1982 "Biophysics of Water". John Wiley & Sons, Ltd. Chichester (in Press).

Fröhlich, H. 1980 Adv. Electron. Electron. Phys. 53, 85-152.

Fulton, A.B. 1982 Cell 30, 345-347.

Gershon, N., K.R. Porter and B. Trus 1982 J. Cell. Biol. 95, 406a.

Hazlewood, C.F. 1979 in: "Cell-Associated Water" (W. Drost-Hansen and J.S. Clegg, eds.) pp. 165-259. Academic Press, N.Y.

Hess, B., A. Boiteux and E.M. Chance 1980 in: "Molecular Biology, Biochemistry and Biophysics" Vol. 32, pp. 157-164 (F. Chapeville and A.L. Haenni, eds.) Springer-Verlag, Berlin.

Heuser, J.E. and M.W. Kirschner 1980 J. Cell Biol. 86, 212-234.

Horowitz, S.B. and T.W. Pearson 181 Molec. Cell Biol. 1, 769-784.

Hunt, R.C. and J.A. Hood 1982 Biochim Biophys. Acta 720, 106-110.

Isenberg, G., K. Leonard and B.M. Jockusch 1982 J. Mol. Biol. 158, 231-249.

Israelachvilli, N.J. 1982 in: "Biophysics of Water" (F. Franks, ed.) John Wiley and Sons, Ltd., Chichester (in Press).

Keith, A. (ed.) 1979 "The Aqueous Cytoplasm" 297 pp. Marcel-Dekkar, N.Y.

Kempner, E.S. 1980 in: "Cell Compartmentation and Metabolic Channeling" (L. Nover, F. Lynen and K. Mothes, eds.). pp. 217-224. Elsevier, Amsterdam.

Korn, E.D. 1982 Physiol. Rev. 62, 672-737.

Kreis, T.E. and W. Birchmeier 1982 Intern. Rev. Cytol. 75, 209-227.

Ling, G.N. 1980 in: "Coopèrative Phenomena in Biology" (G. Karreman, (ed.) pp. 36-99. Pergamon Press, N.Y.

Lis, L.J., M. McAlister, R.P. Rand and V.A. Parsegian 1982 Biophys. J. 37, 657-666.

Mansell, J.L. and J.S. Clegg 1983 Cryobiology, in Press.

Masters, C.J. 1981 "CRC Critical Reviews in Biochemistry" 11, 105-144.

Morton, D.F., J.F. Weidemann and F.M. Clarke 1982 Micron 13, 377-379.

Negendank, W.G. 1982 Biochim. Biophys. Acta 694, 123-161.

Paine, P.L. and S.B. Horowitz 1980 in: "Cell Biology: a Comprehensive Treatise: (L. Goldstein and D. Prescott, eds.) Vol. 4 pp. 299-338. Academic Press, N.Y.

Penman, S., A. Fulton, D. Capo, A. Ben Ze'ev, S. Wittelsberger and C.F. Tse 1982 in: "Cold Spring Harbor Symposia on Quantitative Biology" 46, 1013-1028. Cold Spring Harbor, N.Y.

Porter, K.R. and M.A. McNiven 1982 Cell 29, 23-32.

Poste, G. and G.L. Nicolson (eds.) 1981 "Cytoskeletal Elements and Plasma Membrane Organization" Vol. 7, Cell Surface Reviews 349 pp. Elsevier, Amsterdam.

Prives, J., A.B. Fulton, S. Penman, M.P. Daniels and C.N. Christian 1982 J. Cell Biol. 92, 231-236.

Puca, G.A. and V. Sica 1981 Biochem. Biophys. Res. Com. 103, 682-689.

Rasenick, M.M., P.J. Stein and M.W. Bitensky 1981 Nature, Lond. 294, 560-562.

Sahyoun, N.E., H. LeVine, G.M. Hebdon, R.K. Khouri and P. Cutrecasas 1981 Biochem. Biophys. Res. Com. 101, 1003-1010.

Schliwa, M., J. vanBlerkom and K.R. Porter 1981 Proc. Natl. Acad. Sci. USA 78, 4329-4333.

Sies, H. (ed.) 1982 "Metabolic Compartmentation" 561 pp. Academic Press, N.Y.

Snyderman, R. and E.J. Goetze 1981 Science 213, 830-837.

Starlinger, V.H. 1967 Hoppe-Seyler's Z. Physiol. Chem. 348, 864-872.

Weber, J.P. and S.A. Bernhard 1982 Biochemistry 21, 4189-4194.

Welch, G.R. and T. Keleti 1981 J. Theor. Biol. 93, 701-735.

Welch, G.R., B. Somogyi and S. Damjanovich 1982 Prog. Biophys. Molec. Biol. 39, 109-146.

Westermark, B. and K.R. Porter 1982 J. Cell. Biol. 94, 42-50.

Wheatley, D.N. and M.S. Inglis 1981 Cell Biol. Intern. Rep. 5, 1083-1092.

Coherent Properties of the Membranous Systems of Electron Transport Phosphorylation

DOUGLAS B. KELL and G. DUNCAN HITCHENS

Department of Botany & Microbiology, University College of Wales, Penglais, Aberystwyth, Dyfed, SY23 3DA

1. INTRODUCTION AND SCOPE

"Before entering a house, it is customary to remain a while in the courtyard" - Japanese proverb.

It is now well known that the role of a universal chemical energy currency in living cells is played by the so-called high-energy compound adenosine triphosphate (ATP), whose endergonic synthesis from adenosine diphosphate (ADP) and inorganic phosphate ($\Delta G^{0'}$ = + 31 kJ mol^{-1}) permits the cell to store free energy in a kinetically stable chemical form. One source of the free energy necessary to drive this reaction lies in processes such as oxidative metabolism or photosynthetic electron flow, and the overall process of ATP synthesis coupled to electron transfer is thus referred to as electron transport phosphorylation (see e.g. Stryer, 1981, Lehninger, 1982). The question then arises as to the nature of the free energy transfer between the (exergonic) reactions of electron transport and the otherwise endergonic synthesis of ATP. It is usual to encapsulate this question in the form of a scheme (equation 1) in which a 'high energy intermediate', often denoted "$\sim$" ("squiggle"), constitutes the energetic link between electron transport and ATP synthesis; it is the nature of this "$\sim$" that forms the subject of the present considerations.

$$\text{Electron Transport} \longrightarrow \text{"}\sim\text{"} \longrightarrow \text{ATP} \qquad \text{("}\sim\text{"} \downarrow \text{Heat)} \qquad \text{...... (Eq. 1)}$$

A great conceptual leap forward in the analysis of this problem was made when, in the early 1960's, Mitchell and Williams independently proposed that 'energised' protons might constitute this "$\sim$", and Mitchell's proposals in particular, generally referred to collectively as the chemiosmotic coupling hypothesis, generated a number of successfully tested predictions sufficient to persuade most authorities that the essential mystery of the nature of the "$\sim$" had indeed been solved.

Coherent Excitations in Biological Systems
Ed. by H. Fröhlich and F. Kremer

The central idea of the chemiosmotic coupling hypothesis is that the reactions of electron transport, catalysed by proteinaceous complexes embedded in the so-called coupling membrane, are more or less tightly coupled to the translocation of protons between (phases in equilibrium with) the bulk aqueous phases that the coupling membrane serves to separate. The 'energised intermediate' is then equivalent to the magnitude of the proton electrochemical potential difference $\Delta\tilde{\mu}_{H^+}/F$ or 'protonmotive force' Δp, in electrical units (mV):

$$\Delta p = \Delta\tilde{\mu}_{H^+}/F = \Delta\psi - 2.303RT\,\Delta pH/F \qquad \text{...... (Eq. 2)}$$

This protonmotive force, which, it may be noted, is of a macroscopic thermodynamic character, may act, according to the chemiosmotic model, to drive protons back across the coupling membrane via an H^+-ATP synthase complex. So-called uncoupler molecules, which act to uncouple electron transport from phosphorylation, are taken to act as lipophilic weak acids which catalyse the return of H^+ across the coupling membrane via regions other than the H^+-ATP synthase. These general features are diagrammed in Fig. 1. An excellent introduction to the chemiosmotic coupling hypothesis, and to some of the supporting evidence for this model, may be found in Nicholls's recent lucid monograph (Nicholls 1982), which readers lacking a background in this area are strongly urged to consult.

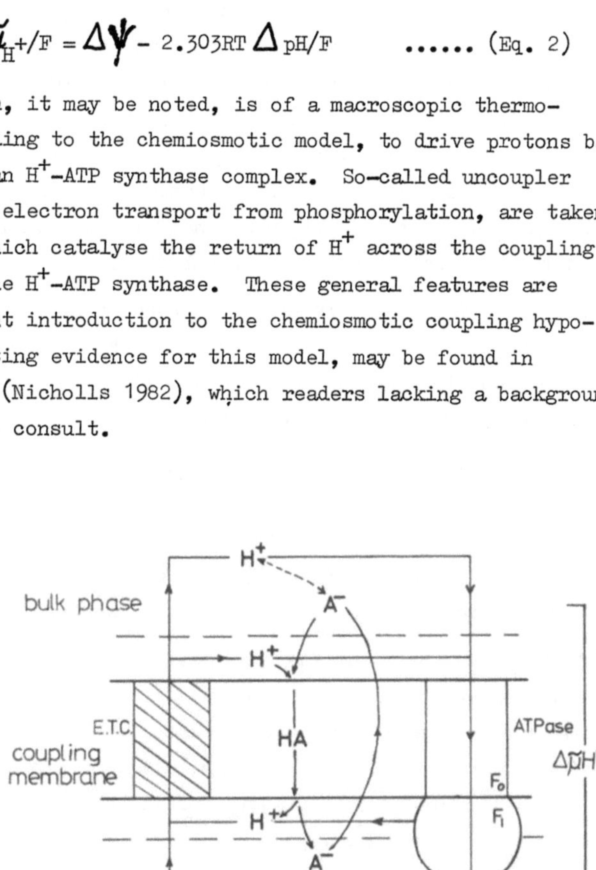

Fig. 1. Energy coupling in protonmotive systems. A coupling membrane containing spatially separate, proteinaceous electron transport (ETC) and F_0F_1-ATP synthase (ATPase) complexes is diagrammed. A protonmotive force ($\Delta\tilde{\mu}_{H^+}$) may be set up, between the bulk aqueous phases that the coupling membrane separates, during electron transport phosphorylation. Other proteins (not shown) may serve to carry the proton current along the membrane surfaces so that the energy coupling H^+ pathway is not in equilibrium with $\Delta\tilde{\mu}_{H^+}$ and the 'high energy intermediate' is constituted not only by protons but by proteins (see text). In either coupling scheme, 'uncouplers' are lipophilic weak acids which can cross the membrane in both charged (A^-) and neutral (HA) forms, thus returning 'energised' H^+ back across the membrane before they can pass to the ATP synthase

It is worth remarking here that this protonmotive force is widely believed to serve as an energetic intermediate between a number of different processes (in addition to electron transport phosphorylation) that are catalysed by various protonmotive complexes in biological membrane systems (see e.g. Skulachev 1980), but we shall here confine our analysis to electron transport phosphorylation, although we recognise that the type of mechanism adopted by Nature for energy and information transfer in this process is likely to be of more general occurrence.

It is fair to state at the outset that the growth of the field of the bioenergetics of protonmotive systems has been such that it has now become both extremely specialised and ramified in nature. For this reason, for limitations of space, and in the spirit of the interdisciplinary scope of this volume, what we shall here attempt is an outline of what we believe to be the salient features of current knowledge and ignorance of the behaviour of protonmotive systems, where we use the word 'protonmotive' to indicate an enzyme complex whose activity is coupled to the transport of protons across the membrane in which it is embedded.

First, we shall briefly allude to studies aimed at the measurement of the magnitude of the protonmotive force as defined above, under various experimental conditions, which lead one to suppose that the simple scheme shown in Eq. 1, with Δp fulfilling the role of "$\sim$", may be inadequate.

Secondly, we shall review and present experimental data on the protonmotive activity of electron transport complexes following short bursts of electron transport, which lead one to conclude that H^+ movements observable in the bulk phase external to the membrane vesicles under study bear only a loose or indirect relationship to the protonmotive activity of various electron transport complexes.

Thirdly, we shall outline the theory and practice of the dual inhibitor titration method, the results of which lead one to take the view that in a number of simple systems the free energy released by a particular electron transport complex is, at least under the described conditions, utilisable only by a particular H^+-ATP synthase, and is not therefore appropriately described as an intermediate with the delocalised, ensemble character of Δp.

Finally, we shall consider the extent to which the systems of electron transport phosphorylation (and related processes) possess the properties expounded in Fröhlich's general theory of coherent excitations and giant dipole oscillations in biology.

2. THE PROTONMOTIVE FORCE: MEASUREMENT AND PROPERTIES

Since the notion of the protonmotive force as a kinetically and thermodynamically competent intermediate in electron transport phosphorylation is generally taken as the central tenet of the chemiosmotic coupling hypothesis, a great deal of experimental work has been directed to the measurement of, and assessment of the role of, this parameter. This work has been authoritatively reviewed several times recently (e.g. Rottenberg 1979a, Ferguson and Sorgato 1982, Schlodder et al, 1982, Azzone et al 1983), and we shall therefore confine ourselves to some summary statements, illustrated by typical examples (c.f. Westerhoff et al 1983a).

(a) when measurements of the protonmotive force are performed using different methods on the same biological system, it is commonly found that the apparent degree of membrane energisation differs greatly from (and usually exceeds) the apparent value of the protonmotive force as determined from the distribution of membrane-permeable ions, acids and bases, particularly in photosynthetic systems (see e.g. Ferguson et al 1979, Elema et al 1978, Clark and Jackson 1981, Baccarini-Melandri et al 1981, Junge, 1982).

(b) rates of electron transport or of phosphorylation are not uniquely dependent upon the magnitude of the apparent protonmotive force measured using a given technique (e.g. Padan and Rottenberg 1973, Baccarini-Melandri et al 1977, 1981, Casadio et al 1978, Sorgato and Ferguson 1979, Kell et al 1978a, Melandri et al 1980, Zoratti et al 1982, Wilson and Forman 1982).

(c) under 'static head' conditions (for definition see e.g. Rottenberg 1978, 1979b, Westerhoff and van Dam 1979), the ratio of the free energy stored as ATP to the magnitude of the apparent protonmotive force seems to vary in an extremely arbitrary fashion (see e.g. Kell et al 1978b,c, Azzone et al 1978, Guffanti et al 1979, 1981, Westerhoff et al 1981, Wilson and Forman 1982).

Such anomalies, in terms of the scheme of equation 1 (in which Δp is taken to constitute the "$\sim$"), have therefore led many to take the view that (i) measurements purporting to estimate the protonmotive force as defined in equation 2 are measuring its magnitude correctly but that this protonmotive force is not the actual 'high-energy intermediate', and/or that (ii) the measurements are correctly measuring the 'high-energy intermediate' but that this is not the protonmotive force as defined in equation 2.

At this point it is worth digressing to take cognizance of the fact that an artificial protonmotive force, applied across the bulk phases that the coupling membrane separates, can induce the synthesis of ATP, although the phosphorylation rate is very finely dependent upon the magnitude of this protonmotive force, and is essentially zero if the applied protonmotive force is below approximately 150 mV (e.g. Thayer and Hinkle 1975, Mills and Mitchell 1982, Schlodder et al 1982). To say the least, therefore, it is not outside the bounds of possibility, and the currently available evidence inclines us to adopt the view, that the protonmotive force as defined in equation 2 does not in fact exceed this 'threshold' value under the usual conditions of electron transport phosphorylation. This assumption would nicely account for all of the foregoing observations.

We may thus enquire as to the nature and properties of the directly observable proton-pumping activity that is in fact catalysed by electron transport complexes.

3. PROTONMOTIVE PROPERTIES OF ELECTRON TRANSPORT CHAINS

Following pioneering experiments with mitochondria (see Mitchell and Moyle 1967), the 'oxygen pulse' method was applied to suspensions of the respiratory microorganism Micrococcus (now Paracoccus) denitrificans by Scholes and Mitchell (1970). In this method (see e.g. Mitchell et al 1979, Reynfarje et al 1979, Wikström and Krab 1980, Nicholls 1982), a pulse of O_2, as air-saturated saline, is added to a well-stirred, weakly-buffered suspension of the membrane vesicles of interest, and the resultant pH changes in the external aqueous phase monitored with a sensitive glass electrode system. The ratio of the number of H^+ translocated across the membrane (extrapolated to the half-time of O_2 reduction) to the number of oxygen atoms added is known as the $\rightarrow H^+/O$ ratio. It is found, in a typical experiment, that the $\rightarrow H^+/O$ ratio is greatly increased in the presence of compounds such as SCN^- or K^+/valinomycin that are believed to cross biological membranes rapidly in a charged form.

According to the conventional chemiosmotic explanation of this behaviour (see e.g. Mitchell 1968, Scholes and Mitchell 1970, Kell 1979, Conover and Azzone 1981, Kell and Morris 1981, Kell and Hitchens 1983a), the relatively low (static) electrical capacitance of the membrane means that the electrically uncompensated transmembrane translocation of a rather small number of H^+ leads to the formation of a large membrane potential (equation 2), which drives H^+ back across the membrane before they may be detected; the method thus relies upon the fact that the magnitude of the pH gradient formed under these conditions (equation 2) is very small. K^+/valinomycin or SCN^- act to dissipate the membrane potential and thus allow all the protons translocated into (or from) the bulk aqueous phase external to the microorganisms to remain there sufficiently long to be measured as a true, limiting stoichiometric $\rightarrow H^+/O$ ratio.

Parenthetically, we may mention that there is a lively current debate concerning the true values of the absolute $\rightarrow H^+/O$ ratios in a number of systems, and the constraints that these values place upon models of the mechanism of protonmotive activity. However, this topic is outside our present scope, and readers may gain an entry to this literature (Wikström and Krab 1980, Wikström et al 1981, Wikström and Penttilä 1982, Nicholls 1982), including that containing our own prejudices (Kell et al 1981b, Vignais et al 1981), elsewhere.

The question to which we wish to address ourselves here concerns the pathway of H^+ that were not seen in the absence of K^+/valinomycin or SCN^-. Did they briefly set up a large protonmotive force (as defined in equation 2), as in the conventional chemiosmotic analysis, or did they never in fact do so? Oxygen-pulse experiments in mitochondria (Archbold et al 1979, Conover and Azzone 1981), in Escherichia coli (Gould and Cramer 1977) and in Paracoccus denitrificans (Hitchens and Kell 1982a, Kell and Hitchens 1983) that have actually sought to answer this question show that the latter proposal must be correct, since (in the absence of K^+/valinomycin or SCN^-)

the (submaximal) $\rightarrow H^+/O$ ratio at low oxygen concentrations is virtually independent of the size of the O_2 pulse. Such experiments have thus been interpreted by the cited authors to indicate that there must be at least 2 types of proton circuits in these membranes, only one of which, not seemingly coupled to phosphorylation (but probably serving in pH homeostasis (Padan et al 1982)), enters the bulk aqueous phase external to the membrane vesicles. This is because if the $\rightarrow H^+/O$ ratio in the absence of K^+/valinomycin or SCN^- is 2.5 (but 7.5 in their presence) when the size of the O_2-pulse is 10 ng atom, then the $\rightarrow H^+/O$ ratio should drop even below 2.5 when the size of the O_2 pulse is doubled, since the membrane potential generated by the reduction of an amount of O_2 corresponding to the smaller pulse should have generated a large enough membrane potential to inhibit further H^+ translocation. Since the observed $\rightarrow H^+/O$ stoichiometry is not decreased (in fact it is unchanged under the conditions described, and according to the values given, in Hitchens and Kell 1982a, Kell and Hitchens 1983), the simple conventional analysis given above must be incorrect.

The use of trains of short (10 μs), saturating light flashes in chromatophores (inverted cytoplasmic membrane vesicles) from photosynthetic bacteria allows not only the number but the frequency of electron transport events to be varied. Thus the 'half-time' of the electron transfer events may be varied whilst determining the stoichiometry of proton translocation, as in the O_2-pulse technique, by a potentiometric system. Such experiments (G.D.H. and D.B.K., unpublished) are displayed in Fig. 2, where it may be seen that, just as is found in O_2-pulse experiments, the number of H^+ observably translocated per flash is increased in the presence of added K^+ and valinomycin. Crucially, this number is the same, for a given reaction mixture, whether the flash rate is 3 Hz or 5 Hz (Fig. 2, 3), indicating that no bulk-phase proton movements have been 'missed'. As discussed above for the case of O_2-pulse experiments, the conventional chemiosmotic analysis ascribes the non-appearance of some of the proton movements under certain conditions to the generation of the macroscopic, delocalised membrane potential component of the protonmotive force (equation 2, and see Fig. 4). However, we again stress that this analysis would require that the $\rightarrow H^+$/flash ratio, when submaximal (see Fig. 3(b)), should be monotonically decreasing as the flash number is raised, a type of behaviour which is not observed (Fig. 3 and see also Cogdell and Crofts 1974). We are again forced to conclude that although macroscopically observable bulk-phase proton movements are coupled to electron transport events, these H^+ are not responsible for feeding back on the electron transport chain to inhibit further proton movements to and from the bulk aqueous phases that the coupling membrane separates. Indeed, it is likely that most, if not all, of these observable H^+ that are translocated across the coupling membrane into a phase in equilibrium with the measuring electrode are translocated essentially in a fashion that is electrically compensated. On this basis, one may surmise that the membrane potential between the bulk phases under these conditions should be very small, as is observed (Vredenberg 1976).

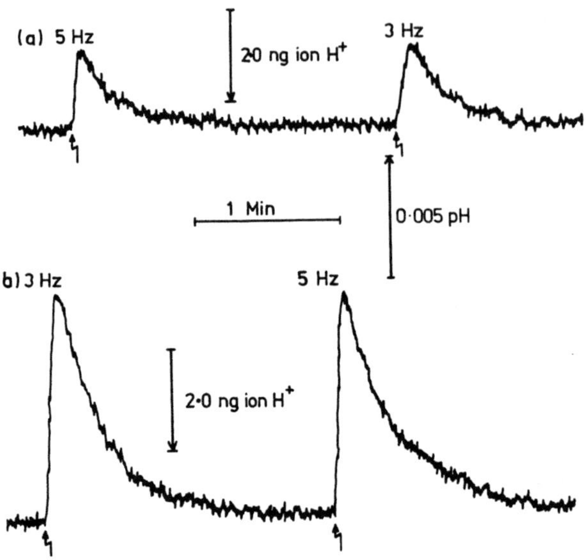

Fig. 2. Light-induced H^+ uptake by Rhodopseudomonas capsulata chromatophores. Chromatophores were prepared as described (Hitchens and Kell 1982b), resuspended and stored in 30 mM K_2SO_4/10 mM $MgSO_4$, adjusted to pH 6.2. Illumination, at the frequencies indicated, was via a PS16 stroboscope (Pro-plan systems, Salisbury; 10 μs flashes, 2.1J/flash). Light was filtered, and H^+ movements were measured in a stirred 6 ml reaction cell, as described (ibidem). Bacteriochlorophyll concentration was 10 μM, and control experiments ensured that the illumination was saturating. Signals from the pH electrode were amplified by an ISEAMP 200 (bandwidth 15 Hz) designed and constructed by Dulas Engineering, Llwyngwern Quarry, Machynlleth. 20 flashes were given in each case at the points arrowed. In (a) 100 mM KCl was also present, whilst in (b) valinomycin (1 μg/ml) was further added. No H^+ movements were seen in the presence of 1 μM carbonyl cyanide p-trifluoromethoxy phenylhydrazone (not shown)

Thus the conclusion from the foregoing type of experiment is that the observable protonmotive events measured using conventional techniques are not adequately accommodated in a simple scheme such as that of Equation 1 and in which "$\sim$" is represented by Δp. For reasons such as those alluded to in the foregoing, therefore, we and others have sought an experimental approach to the problem of electron transport phosphorylation that is independent of the measurement either of bulk-phase proton movements or of the protonmotive force as defined in equation 2. One such approach forms the subject of the following section.

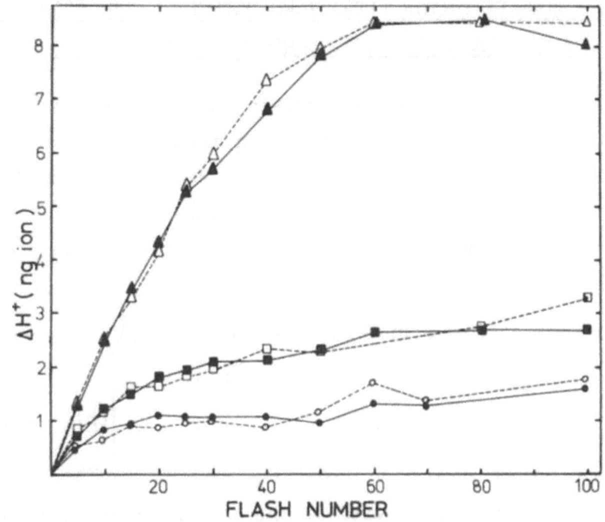

Fig. 3. Effect of KCl, valinomycin and flash number on light-induced H^+ uptake by Rps. capsulata chromatophores. Light-induced H^+ uptake was measured as described in the legend to Fig. 2 at frequencies of 3 Hz (open symbols) and 5 Hz (closed symbols). The basic reaction medium (30 mM K_2SO_4/10 mM $MgSO_4$) (○ , ●) was supplemented with 100 mM KCl (□ , ■) or with 100 mM KCl plus valinomycin (1 μg/ml) (△ , ▲) as indicated, and the figure includes the data of Fig. 2

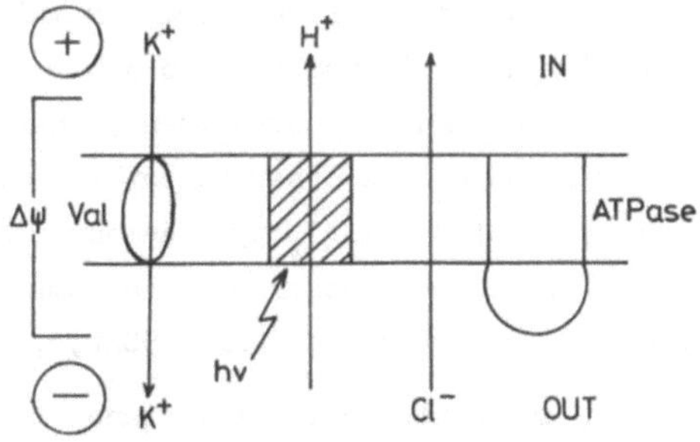

Fig. 4. Conventional chemiosmotic explanation of light-induced H^+ uptake by bacterial chromatophores. The protonmotive segment of the photosynthetic electron transport chain (shaded) pumps a proton from the external bulk phase which sets up a delocalised membrane potential $\Delta\psi$, which acts to inhibit further protonmotive activity on subsequent flashes. Uniport of Cl^- or K^+ (in the presence of valinomycin) decrease $\Delta\psi$ and increase the $\rightarrow H^+$/flash ratio. For further discussion, see text

4. DUAL-INHIBITOR TITRATIONS DETERMINE THE DEGREE OF COHERENCE BETWEEN INDIVIDUAL MOLECULAR STEPS IN A SEQUENCE OF REACTIONS.

The idea behind the dual-inhibitor titration approach was first expounded, at least in the context of electron transport phosphorylation, by Baum and colleagues (Baum *et al* 1971, Baum 1978), and may be expressed as follows. If we have available tight-binding and specific inhibitors of electron transport and of the H^+-ATP synthase, we may modify the delocalised chemiosmotic scheme conforming to that of equation 1 thus:

$$\text{Electron transport} \xrightarrow[I_1]{} \Delta p \xrightarrow[I_2]{} \text{ATP} \qquad \text{...... Eq 3}$$

$$\Delta p \xrightarrow{\oplus \leftarrow \text{uncoupler}} \text{heat}$$

so that the sensitivity of the overall pathway flux (rate of phosphorylation) to I_1 (electron transport inhibitor) will be decreased by the presence of a partially inhibitory titre of I_2 (H^+-ATP synthase inhibitor) if Δp as defined in equation 2 does serve as a delocalised coupling intermediate. By 'delocalised', we mean the concept that any free energy released (say as Δp) by a particular electron transport chain is available to any H^+-ATP synthase in the membrane vesicle. If energy coupling (or for that matter a metabolic pathway) is microscopic in nature, in the sense that quanta of free energy or material interacting with individual enzymes in the system are not available to other enzyme molecules of the same type in contact with the same aqueous compartment, then the presence of I_2 will have no effect upon the titration of the observed pathway flux with I_1 and *vice versa*. In practice, experiments of the I_1/I_2 type demonstrate, at first sight unequivocally, that the latter behaviour is observed in a number of systems (Fig. 5; Baum *et al* 1971, Baum 1978, Hitchens and Kell, 1982b,c, Westerhoff *et al* 1982a,b, 1983a,b).

Now, as pointed out for instance by Parsonage and Ferguson (1982), such behaviour could also be accommodated in a delocalised, chemiosmotic coupling scheme of the type shown in equations 1 and 3 if the initial portions of the two titration curves were accompanied by a decrease (of the same magnitude) in the value of the protonmotive force. However, a variety of measurements of the protonmotive force indicate that even quite severe inhibition of electron transport by an I_1-type inhibitor do not decrease the protonmotive force (e.g. Kell et al 1978a, Sorgato and Ferguson 1979), in particular when bacterial chromatophores are titrated with antimycin A (Venturoli and Melandri 1982) as in Fig. 5. Wikström (see Westerhoff *et al* 1983b) has raised the interesting possibility that the binding of such inhibitors might be an energy-dependent process (i.e. dependent in some fashion upon the magnitude of the steady-state Δp), and that this behaviour (if true) could lead to an artefactual

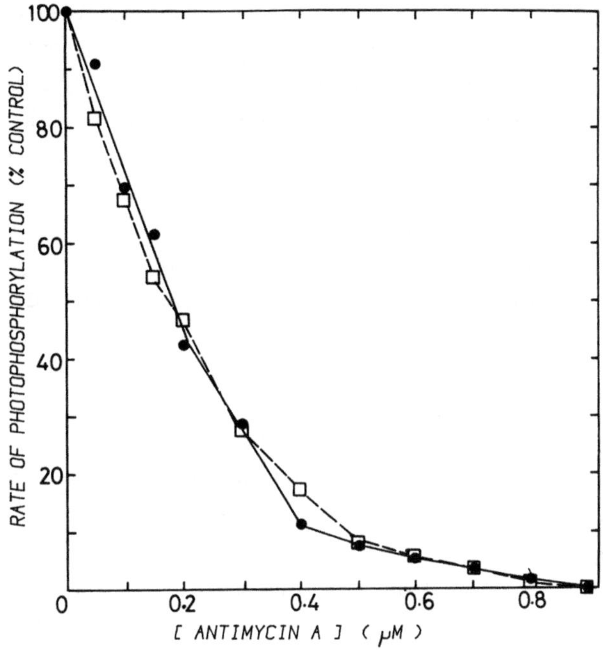

Fig. 5. Effect of dicyclohexyl carbodimide (DCCD) and antimycin A on phosphorylation by Rps. capsulata chromatophores. The rate of phosphorylation was titrated with the electron transport inhibitor antimycin A (●, □). In one case (□) chromatophores were preincubated with a concentration (50 μM) of the covalent H^+-ATP synthase inhibitor DCCD sufficient to reduce the rate of phosphorylation in the absence of antimycin by 50%. Control rates were 1.18 and 0.58 μmol ATP/min/mg bacteriochlorophyll. Were energy couplings effected via a delocalised intermediate, such as Δp, the second titration (□) should initially have had a lesser slope than the first; it is evident that the 2 titration curves are virtually identical. Data after Hitchens and Kell (1982b)

interpretation of this type of experiment. At least two lines of reasoning serve to exclude this criticism: (a) I_1/I_2 titrations are symmetrical, in the sense that neither of the pair decreases the inhibitory properties of the other (see e.g. Hitchens and Kell 1982c), and since the inhibitors act on different sides of the 'high energy intermediate' they would be expected to behave differently depending upon which way the experiment is performed (i.e. which inhibitor concentration forms the abscissa in experiments of the type in Fig. 5) if energy-dependent binding effects might be of importance to the analysis; (b) the use of a covalent H^+-ATP synthase inhibitor in concert with trains of saturating light flashes of different frequency (i.e. conditions in which no energy-dependent inhibitor binding may be invoked) demonstrates that the same type of localised energy coupling is observed in bacterial chromatophores (Venturoli and Melandri 1982).

It should be stressed that the number of ATP molecules synthesised per pair of electrons passing down an electron transport chain is, for a well-coupled system as isolated, in the absence of added uncoupler molecules, independent of the rate of electron transport over a fairly wide range (van Dam and Tsou 1970, Ernster and Nordenbrand 1974, Ferguson et al 1976, Graber and Witt 1976, Jackson et al 1981). This shows that any macroscopic native 'energy leak' in the system, loosely shown as 'heat' in the schemes of equations 1 and 3, is insignificant under these conditions. Further, an I_2-type inhibitor cannot be expected to lower the value of a delocalised intermediate such as Δp. Thus, whatever the putative relationship between Δp and the rate of phosphorylation in a scheme such as that of equation 3, the titre of uncoupler required for full uncoupling cannot be decreased by an I_2-type inhibitor, since equation 3 supposes that uncouplers act only by decreasing Δp. The remarkable fact is, however, that a variety of uncouplers and ionophores act more potently in bacterial chromatophores (Hitchens and Kell 1982c, 1983a,b) and in submitochondrial particles (Westerhoff et al 1982a, 1983a,b) when the reaction supposedly driven by Δp is partially restricted by an I_2-type inhibitor. In terms of the 'localised coupling' alluded to above, however, this result is indeed to be expected, since the concept of localised coupling implies, in this context, that inhibition of an H^+-ATP synthase molecule abolishes the utilisability elsewhere of a quantum of free energy from 'its' electron transport chain complexes. Thus uncouplers only seek out coupling units in which the whole unit is potentially active, and will work better when this number is decreased by an I_2-type inhibitor, as observed. Thus the free energy transfer is not, in this type of system, a stochastic process.

It is germane to point out here that the foregoing, rather qualitative, expose has a quantitative counterpart, often termed 'metabolic control theory', that has been applied to metabolic pathways (most recently reviewed in Groen et al 1982). This latter formalism is itself a 'delocalised' (macroscopic) one in the sense alluded to above, and, in terms of this latter formalism, it would be found experimentally that the sum of the 'control strengths' of the individual steps of electron transport phosphorylation in the experiments alluded to here exceeds 1. As pointed out (Groen et al 1982), such a general finding poses formidable problems for metabolic control theories of this type, and, we would aver, provides a powerful experimental approach to the testing of the important concepts of a rather more sophisticated cellular organisation than those that are commonly held (see e.g. Welch 1977, Kell 1979, Berry 1981, Clegg 1981 and this volume; Welch and Berry, this volume).

In summary, we may state that, although even more subtle and arcane interactions almost certainly take place in energy coupling membranes (Kell et al 1981a), the approach outlined in this section shows that in a number of systems the energy coupling events within the separate molecular complexes of electron transport phosphorylation may be said to possess the property of coherence in that the successful transfer of free energy between electron trnnsport and H^+-ATP synthase complexes occurs strictly at the level of the individual molecular complexes. Thus free energy can only be transferred by a particular electron transport complex in a form suitable for driving the synthesis of ATP under the conditions described if a particular H^+-ATP synthase is potentially active. We shall now explore this concept in more detail.

5. COHERENCE IN THE PROCESS OF ELECTRON TRANSPORT PHOSPHORYLATION

5.1. The Fröhlich theory of long-range coherent oscillations in biology

Over the last fifteen years or so, H. Fröhlich has developed, and stimulated experimental study upon, a theory of long-range coherent excitations in biology that we think is likely to be of the greatest relevance to the problem of free-energy transfer in electron transport phosphorylation. Since a recent review is available (Fröhlich 1980), we shall discuss only what, according to our reading of the model, seem to be its salient features for our present purposes:

(a) one or more polar modes of the constituents of biomembranes may be coherently excited through energy supplied from metabolic processes, provided that the rate of energy supply exceeds a critical, threshold value; the oscillations are then condensed into a single, lower frequency mode, a phase transition analogous to the Einstein condensation of a Bose gas and to the many other types of phase transition summarised by Haken (1977);

(b) such states may be stabilised by non-linear interactions between the electrical and vibrational modes of the charged membranes and its heat bath, where the word 'membrane' is taken to include both the lipid and protein constituents and the adsorbed ions and water molecules. Under appropriate conditions, this mode possesses the properties of a relatively long-lived (metastable) ferroelectric state (Bilz et al 1981, Kaiser 1981);

(c) externally applied electromagnetic energy may feed into the system to raise it into, or destroy, the metastable state;

(d) free energy may be transferred over relatively long distances as a wave of polarisation (Fröhlich 1968). The ferroelectric state should be experimentally observable by dielectric measurements (Fröhlich 1975) as a giant, oscillating dipole.

We shall therefore discuss these points, from a bioenergeticist's standpoint, in terms of what is known of the process of electron transport phosphorylation, and may begin by remarking upon the striking similarities between the Fröhlich model and those invoked by us, for quite different reasons, in this context (Kell et al 1981a, Kell and Morris 1981).

5.2. 'Metabolic pumping' by protonmotive systems

It is now axiomatic that a variety of membrane-located metabolic enzymes are protonmotive, as discussed above; indeed, one is inclined to remark on the fortuitous semantic coincidence in the use of the term 'pump' by the physicists and biologists in this context. As discussed above, therefore, both the requirement for a threshold value of the free energy supply and the pumping of protons across the dielectric barrier of coupling membranes are well-recognized features of the process of electron transport phosphorylation.

5.3. The existence of non-linear interactions in energy coupling membranes

As discussed in section 4, it is an experimental observation that individual, spatially separate redox and ATP synthase proton pumps exhibit a very strict coupling relationship. This of itself requires a highly non-linear interaction. The electrical and elastic properties of biomembranes are not usually considered from the standpoint of the type of idea explicit in the Fröhlich model, though recent reviews that perhaps most closely approach this analysis may be consulted (Evans and Skalak 1980, Miller 1981). From the bioenergetic viewpoint, one of the particularly important virtues of the Fröhlich model is the demonstration that such non-linear interactions leading to a stabilisation of one or more coherent modes may exist, and should perforce be directly sought experimentally.

5.4. Absorption and emission of electromagnetic energy by protonmotive systems

One of the especially noteworthy features of the Fröhlich model is the prediction that very weak electromagnetic radiation of frequency ca. 100 GHz may be expected to change the properties of the types of system under discussion by purely non-thermal means. A number of very striking observations (e.g. Grundler and Keilmann 1978, Grundler et al, Kremer, Nimtz , this volume) have shown that this behaviour can indeed be observed in intact living cells. The suggestions concerning the importance of the above (microwave) frequency range were prompted by a consideration of the relationship between the thickness of biomembranes (10^{-8}m) and the likely speed of sound (phonons) therein (1000 ms^{-1}). However (Fröhlich 1980), many types of factor may serve to lower the frequency range in which weak external electromagnetic fields may affect biomembrane systems. One should of course note, in this context, the fact that large external electric field pulses may in fact be used to drive ATP synthesis, just as may large pH jumps (e.g. Teissie et al 1981, Gräber et al 1982, Hamamoto et al 1982, Schlodder et al 1982) although the exact mechanisms at work remain far from clear (Vinkler and Korenstein 1981, Vinkler et al 1982). Weak microwaves have been reported not to affect oxidative phosphorylation in isolated rat-liver mitochondria (Elder and Ali 1975), but irradiation for 15 minutes at 20 kHz did seem to inhibit this process by about 20% (Straub and Carver 1975). Since any such effects are expected to be highly frequency-dependent, a more systematic study than seems so far to have been attempted, using isolated protonmotive systems, seems warranted. Similarly, it is not possible for us at this stage to determine whether the very exciting observations of enhanced anti-Stokes Raman scattering from a number of active microorganisms (e.g. Drissler and Macfarlane 1978, Webb 1980, Del Guidice et al 1982, Drissler, this volume) may be due to energised membrane states involved in electron transport phosphorylation. The emission of electromagnetic energy from a number of cells that may be detected by microdielectrophoresis (Pohl 1981 and this volume) or by the vibrating probe method (Jaffe 1981) does, however, seem to be best viewed as a membrane phenomenon.

As indicated above, perhaps the least equivocal technique that might be applied, in the present context, to biological systems generally, and protonmotive systems in particular, is to compare their dielectric properties under active (working) and inactive (dormant or equilibrium) conditions (Fröhlich 1975). Unfortunately, we are not aware of any experiments that have so far sought differences in the dielectric properties of biological systems in different, identified metabolic or catalytic states. This current experimental lacuna may be ascribed to the lack, until recently, of instrumentation of a rapidity adequate to measure these properties accurately under steady-state, non-equilibrium conditions. However, the theories and data available to date on dormant or equilibrium systems do enable us to introduce some considerations that will permit us to adopt the optimistic conclusion that the extension of dielectric spectroscopy to working biochemical systems, especially protonmotive ones, is likely to prove of value. Since high-frequency work is discussed elsewhere (Genzel, this volume) we confine ourselves to work below 100 MHz.

5.5. Conformational dynamics and the dielectric properties of biochemical systems

In the case of enzymes, including those of electron transport phosphorylation, which typically have a turnover number in the range 1-10 ms, a plethora of experimental approaches have demonstrated that isolated proteins exhibit a variety of conformational fluctuations in a frequency range of ca. 10^{-5}Hz to 10^{12} Hz. The view has evolved, therefore, that an isolated, substrateless enzyme is, in Weber's (1975) evocative phrase, 'a kicking and screaming stochastic molecule'.

We may then ask the question "to what extent do the high frequency motions of active and inactive membranes and proteins bear a causal relationship to each other and to those of lower frequency?"; two marvellous recent reviews (Careri et al 1979, Welch et al 1981) have summarised the idea that a correlational or causal relationship between the various fluctuational properties of proteins is indeed crucial to enzymatic function. Are these properties perhaps observable by dielectric spectroscopy in proteins generally? Although, we may reiterate, changes in dielectric properties during enzyme activity have not so far been sought successfully, the available evidence is usually interpreted to favour two main mechanisms causing dielectric dispersions in isolated (inert) globular proteins (e.g. Takashima and Minakata 1975, Grant et al

1978, Pethig 1979, Gascoyne et al 1981 and references therein; see also Hasted, this volume):

(a) Debye-like rotation of the entire protein molecule due to its possession of a permanent dipole moment;

(b) ion, and especially proton, movement to and from, and between, groups on the protein surface. (We do not here discuss the relaxation of protein-bound water).

Actually, the overwhelming majority of models assume an independence from each other of the elementary processes leading to the observed, macroscopic dielectric dispersions, but, as indicated above, there are reasons to suppose that this is likely to turn out to be something of an oversimplification. The fit between theory and experiment found in even the deepest studies (e.g. South and Grant 1972, Petersen and Cone 1975) leaves abundant scope to invoke factors other than rotation of the whole protein or protonic fluctuations. We would suggest, therefore, that correlated conformational fluctuations in different regions of the tertiary structure of the protein backbone itself may constitute an additional important mechanism of dielectric dispersion (c.f. White and Slutsky 1972). Formally, they would appear as a non-additive superposition of the small changes in (static and mean square) dipole moment caused by the individual conformational fluctuations of charged and polar groups of interest.

As far as non-metabolising, membrane-bound cells and vesicles are concerned (e.g. Schwan 1957, Schanne and Ceretti 1978, Pethig 1979, Stoy et al 1982 and references therein), a large body of dielectric measurements by a number of workers indicates that the presence of the membrane gives rise to two main dispersions; the β-dispersion occurs in the radio-frequency range, and is mostly ascribable to a Maxwell-Wagner effect at the interface between the aqueous phases and the poorly-conducting cytoplasmic membrane. Interestingly, however, concentrations of uncoupler sufficient to release maximally the respiratory control of intact *Paracoccus denitrificans* cells have only a marginal effect on the β-dispersion in this organism (G.D.H. and D.B.K., in preparation). The α-dispersion is seen in the audio-frequency range, and seems to be dominated by the relaxation of bound ions tangential to charged surfaces (Schwarz 1962, Dukhin and Shilov 1974).

Since, from the earlier discussion, we may expect 'energised' coupling membranes to possess non-equilibrium surface charge distributions, we may anticipate striking, and possibly resonant, dielectric properties to be observed in the audio-frequency range under appropriate conditions. Such a search is in progress.

In summary, the available dielectric data on biological systems in states of electrochemical equilibrium allow one to conclude that if the types of electrical states envisaged in the Fröhlich model are formed during electron transport phosphorylation, they should indeed be amenable to experimental observation by dielectric spectroscopy, as suggested (Fröhlich 1975).

6. CONCLUDING REMARKS

The foregoing outline of electron transport phosphorylation may be summarised briefly as follows:

(a) there is abundant evidence that proteinaceous complexes in electron transport phosphorylation are protonmotive;

(b) it is not easy to understand their properties in terms of macroscopic coupling theories which invoke a delocalised proton electrochemical potential (or other delocalised, macroscopic thermodynamic force) as 'the' intermediate of electron transport phosphorylation;

(c) the conception of the 'energised state' of coupling membranes as a giant oscillating dipole with non-linearly coupled electrical and vibrational modes, as described in Fröhlich's theory of long-range interactions in biology, is consistent with the available experimental data on these systems, and suggests some extremely interesting experimental tests of, in bioenergetic terms, a novel character.

7. ACKNOWLEDGEMENTS

We thank the Science and Engineering Research Council for financial support. We acknowledge with pleasure many stimulating discussions with Dr Hans Westerhoff, and the receipt of material prior to publication. We are indebted to Professor H. Fröhlich for many clarifying, stimulating and thought-provoking discussions.

8. REFERENCES

ARCHBOLD, G.P.R, FARRINGTON, C.L., LAPPIN, S.A., McKAY, A.M. and MALPRESS, F.M. (1979) Biochem. J. 180, 161-174.

AZZONE, G.F., PIETROBON, D. and ZORATTI, M. (1983) Curr. Top. Bionerg., in press.

AZZONE, G.F., POZZAN, T. and MASSARI, S. (1978) Biochim. Biophys. Acta 501, 307-316.

BACCARINI-MELANDRI, A., CASADIO, R. and MELANDRI, B.A. (1977) Eur. J. Biochem. 78, 389-402.

BACCARINI-MELANDRI, A., CASADIO, R. and MELANDRI, B.A. (1981) Curr. Top. Bioenerg., 12, 197-258.

BAUM, H. (1978) in 'The Molecular Biology of Membranes' (Fleischer, S., Hatefi, Y., MacLennan, D.H. and Tzagoloff, A., eds.) pp. 243-262. Plenum, New York.

BAUM, H., HALL, G.S., NALDER, J. and BEECHEY, R. . (1971) in 'Energy Transduction in Respiration and Photosynthesis' (Quagliariello, E., Papa, S. and Rossi, C.S. eds) pp. 747-755. Adriatica Editrice, Bari.

BERRY, M.N. (1981) FEBS Lett. 134, 133-138.

BILZ, H., BÜTTNER, H. and FRÖHLICH, H. (1981) Z. Naturforsch. 36b, 208-212.

CARERI, G., FASELLA, P. and GRATTON, E. (1979) Ann. Rev. Biophys. Bioeng. 8, 69-97.

CASADIO, R., BACCARINI-MELANDRI, A. and MELANDRI, B.A. (1978) FEBS Lett. 87, 323-328.

CLARK, A.J. and JACKSON, J.B. (1981) Biochem. J. 200, 389-397.

CLEGG, J.S. (1981) Collective Phenomena 3, 289-312.

COGBELL, R.J. and CROFTS, A.R. (1974) Biochim. Biophys. Acta 347, 264-272.

CONOVER, T.E. and AZZONE, G.F. (1981) in 'Mitochondria and Microsomes' (Lee, C-P., Schatz, G. and Dollner, G. eds.) pp. 481-518. Addison-Wesley, New York.

DEL GUIDICE, E., DOGLIA, S. and MILANI, M. (1982) Phys. Scripta 26, 232-238.

DRISSLER, F. and MACFARLANE, R.M. (1978) Phys. Lett. 69A, 65-67.

DUKHIN, S.S. and SHILOV, V.N. (1974) 'Dielectric Phenomena and the double layer in disperse systems and polyelectrolytes' Wiley, Chichester.

ELDER, J.A. and ALI, J.S. (1975) Ann. N.Y. Acad. Sci. 247, 251-262.

ELEMA, R.P., MICHELS, P.A.M. and KONINGS, W.N. (1978) Eur. J. Biochem. 92, 381-387.

ERNSTER, L. and NORDENBRAND, K. (1974) BBA Library 13, 283-288.

EVANS, E.A. and SKALAK, R. (1980) 'Mechanics and Thermodynamics of Biomembranes'. CRC Press, Boca Raton.

FERGUSON, S.J., JOHN, P., LLOYD, W.J., RADDA, G.K. and WHATLEY, F.R. (1976) FEBS Lett. 62, 272-275.

FERGUSON, S.J., JONES, O.T.G., KELL, D.B. and SORGATO, M.C. (1979) Biochem. J. 180, 75-85.

FERGUSON, S.J. and SORGATO, M.C. (1982) Ann. Rev. Biochem. 51, 185-217.

FRÖHLICH, H. (1968) Nature 219, 743-744.

FRÖHLICH, H. (1975) Proc. Natl. Acad. Sci. 72, 4211-4215.

FRÖHLICH, H. (1980) Adv. Electron. Electron Phys. 53, 85-152.

GASCOYNE, P.R.C., PETHIG, R. and SZENT-GYÖRGYI, A. (1981) Proc. Natl. Acd. Sci. 78, 261-265.

GOULD, J.M. and CRAMER, W.A. (1977) J. Biol. Chem. 252, 5875-5882.

GRÄBER, P., ROGNER, M. and BUCHWALD, H-E.(1982) FEBS Lett. 145, 35-40.

GRÄBER, P. and WITT, H.T. (1976) Biochim. Biophys. Acta 423, 141-163.

GRANT, E.H., SHEPPARD, R.J. and SOUTH, G.P. (1978) 'Dielectric behaviour of biological molecules in solution'. Clarendon Press, Oxford.

GROEN, A.K., van der MEER, R., WESTERHOFF, H.V., WANDERS, R.J.A., AKERBROOM, T.P.M. and TAGER, J.M. (1982) in 'Metabolic Compartmentation' (Sies, H. ed.) pp. 9-37. Academic, New York.

GRUNDLER, W. and KEILMANN, F. (1978) Z. Naturforsch. 33c, 15-22.

GUFFANTI, A.A., BLUMENFELD, H.H. and KRULWICH, T.A. (1981) J. Biol. Chem. 256, 8418-8421.

GUFFANTI, A.A., SUSMAN, P., BLANCO, R. and KRULWICH, T.A. (1978) J. Biol. Chem. 253, 708-715.

HAKEN, H. (1977) 'Synergetics'. Springer, Heidelberg.

HAMAMOTO, T., OHNO, K. and KAGAWA, Y. (1982) J. Biochem. 91, 1759-1766.

HITCHENS, G.D. and KELL, D.B. (1982a) Biochem. Soc. Trans. 10, 261.

HITCHENS, G.D. and KELL, D.B. (1982b) Biochem. J. 206, 351-357.

HITCHENS, G.D. and KELL, D.B. (1982c) Biosci. Rep. 2, 743-749.

HITCHENS, G.D. and KELL, D.B. (1983a) Biochem. J. 210, in press.

HITCHENS, G.D. and KELL, D.B. (1983b) Biochim. Biophys. Acta, submitted.

JACKSON, J.B., VENTUROLI, G., BACCARINI-MELANDRI, A. and MELANDRI, B.A. (1981) Biochim. Biophys. Acta 636, 1-8.

JAFFE, L.F. (1981) Phil. Trans. R. Soc. B 295, 553-566.

JUNGE, W. (1982) Curr. Top. Membr. Trans. 16, 431-465.

KAISER, F. (1981) ACS Symp. Ser. 157, 219-241.

KELL, D.B. (1979) Biochim. Biophys. Acta 549, 55-99.

KELL, D.B., FERGUSON, S.J. and JOHN, P. (1978a) Biochem. Soc. Trans. 6, 1292-1295.

KELL, D.B., JOHN, P. and FERGUSON, S.J. (1978b) Biochem. J. 174, 257-266.

KELL, D.B., FERGUSON, S.J. and JOHN, P. (1978c) Biochim. Biophys. Acta 502, 111-126.

KELL, D.B., CLARKE, D.J. and MORRIS, J.G. (1981) FEMS Microbiol. Lett. 11, 1-11.

KELL, D.B., CLARKE, D.J., BURNS, A. and MORRIS, J.G. (1981b) Specul. Sci. Technol. 4, 109-120.

KELL, D.B. and HITCHENS, G.D. (1983) Disc. Faraday Soc. 74, in press.

KELL, D.B. and MORRIS, J.G. (1981) in 'Vectorial Reactions in Electron and Ion Transport in Mitochondria and Bacteria' (Palmieri, F., Quagliariello, E., Siliprandi, N. and Slater, E.C., eds.) pp. 339-347. Elsevier, Amsterdam.

LEHNINGER, A.L. (1982) 'Principles of Biochemistry' Worth, New York.

MELANDRI, B.A., VENTUROLI, G., de SANTIS, A. and BACCARINI-MELANDRI, A. (1980) Biochim. Biophys. Acta 592, 38-52.

MILLER, I.R. (1981) in 'Topics in Bioelectrochemistry and Bioenergetics' (Milazzo, G. ed.) Vol. 4, pp. 162-224. Wiley, London.

MILLS, J.D. and MITCHELL, P. (1982) FEBS Lett. 144, 63-67.

MITCHELL, P. (1968) 'Chemiosmotic coupling and energy transduction'. Glynn Research, Bodmin.

MITCHELL, P. and MOYLE, J. (1967) Biochem. J. 105, 1147-1162.

MITCHELL, P., MOYLE, J. and MITCHELL, R. (1979) Meth. Enzymol. 55, 627-640.

NICHOLLS, D.G. (1982) 'Bioenergetics. An Introduction to the Chemiosmotic Theory'. Academic Press, London.

PADAN, E. and ROTTENBERG, H. (1973) Eur. J. Biochem. 40, 431-437.

PADAN, E., ZILBERSTEIN, D. and SCHULDINER, S. (1981) Biochim. Biophys. Acta 650, 151-166.

PARSONAGE, D. and FERGUSON, S.J. (1982) Biochem. Soc. Trans. 10, 257-258.

PETERSEN, D.C. and CONE, R.A. (1975) Biophys. J. 15, 1181-1200.

PETHIG, R. (1979) 'Dielectric and Electronic Properties of Biological Materials'. Wiley, Chichester.

POHL, H.A. (1981) Collective Phenomena 3, 221-244.

REYNAFARJE, B., BRAND, M.D., ALEXANDRE, A. and LEHNINGER, A.L. (1979) Meth. Enzymol. 55, 640-656.

ROTTENBERG, H. (1978) Progr. Surf. Membr. Sci. 12, 245-325.

ROTTENBERG, H. (1979a) Meth. Enzymol. 55, 547-569.

ROTTENBERG, H. (1979b) Biochim. Biophys. Acta 549, 225-253.

SCHANNE, O.F. and CERETTI, E.R.P. (1978) 'Impedance Measurements in Biological Cells'. Wiley, Chichester.

SCHLODDER, E., GRÄBER, P. and WITT, H.T. (1982) in 'Electron Transport and Photophosphorylation' (Barber, J. ed.) pp. 105-175. Elsevier, Amsterdam.

SCHOLES, P. and MITCHELL, P. (1970) J. Bioenerg. 1, 309-323.

SCHWAN, H.P. (1957) Adv. Biol. Med. Phys. 5, 147-209.

SCHWARZ, G. (1962) J. Phys. Chem. 66, 2636-2642.

SKULACHEV, V.P. (1980) Can. J. Biochem. 58, 161-175.

SORGATO, M.C. and FERGUSON, S.J. (1979) Biochemistry 18, 5737-5742.

SOUTH, G.P. and GRANT, E.H. (1972) Proc. R. Soc. A 328, 371-387.

STOY, R.D., FOSTER, K.R. and SCHWAN, H.P. (1982) Phys. Med. Biol. 27, 501-513.

STRAUB, K.D. and CARVER, P. (1975) Ann. N.Y. Acad. Sci. 247, 292-300.

STRYER, L. (1981) 'Biochemistry, 2nd Ed.' Freeman, San Francisco.

TAKASHIMA, S. and MINAKATA, A. (1975) in 'Digest of Literature on Dielectrics' (Vaughan, A. ed.) pp. 602-653. National Research Council, Washington.

TEISSIE, J., KNOX, B.E., TSONG, T.Y. and WEHRLE, J. (1981) Proc. Natl. Acad. Sci. 78, 7473-7477.

THAYER, W.S. and HINKLE, P.C. (1975) J. Biol. Chem. 250, 5330-5335.

van DAM, K. and TSOU, C.S. (1970) in 'Electron Transport and Energy Conservation' (Tager, J.M., Papa, S., Quagliariello, E. and Slater, E.C. eds) pp. 421-425. Adriatica Editrice, Bari.

VENTUROLI, G. and MELANDRI, B.A. (1982) Biochim. Biophys. Acta 682, 8-16.

VIGNAIS, P.M., HENRY, M-F., SIM, E. and KELL, D.B. (1981) Curr. Top. Bioenerg. 12, 115-196.

VINKLER, C. and KORENSTEIN, R. (1982) Proc. Natl. Acad. Sci. 79, 3183-3187.

VINKLER, C., KORENSTEIN, R. and FARKAS, D.L. (1982) FEBS Lett. 145, 235-240.

VREDENBERG, W.J. (1976) in 'The Intact Chloroplast' (Barber, J. ed.) pp. 53-88. Elsevier, Amsterdam.

WEBB, S.J. (1980) Phys. Rep. 60, 201-224.

WEBER, G. (1975) Adv. Prot. Chem. 29, 1-83.

WELCH, G.R. (1977) Progr. Biophys. Mol. Biol. 32, 103-191.

WELCH, G.R., SOMOGYI, B. and DAMJANOVICH, S. (1982) Progr. Biophys. Mol. Biol. 39, 109-146.

WESTERHOFF, H.V. and van DAM, K. (1979) Curr. Top. Bioenerg. 9, 1-62.

WESTERHOFF, H.V., SIMONETTI, A.L.M. and van DAM, K. (1981) Biochem. J. 200, 193-202.

WESTERHOFF, H.V., de JONGE, P.C, COLEN, A., GROEN, A.K., WANDERS, R.J.A., van den BERG, G.B. and van DAM, K. (1982a) EBEC Reports 2, 267-268.

WESTERHOFF, H.V., COLEN, A. and van DAM, K. (1982b) Biochem. Soc. Trans., in press.

WESTERHOFF, H.V., MELANDRI, B.A., VENTUROLI, G. and KELL, D.B. (1983a) FEBS Lett., to be submitted.

WESTERHOFF, H.V., HELGERSON, S.L., THEG, S.M., van KOOTEN, O., WIKSTRÖM, M. et al (1983b) Acta Biol. Acad. Sci. Hung., in press.

WHITE, R.D. and SLUTSKY, L.J. (1972) Biopolymers 11, 1973-1984.

WIKSTRÖM, M.K.F. and KRAB, K. (1980) Curr. Top. Bioenerg. 10, 51-101.

WIKSTRÖM, M., KRAB, K. and SARASTE, M. (1981) Ann. Rev. Biochem. 50, 623-655.

WIKSTRÖM, M. and PENTTILÄ, T. (1982) FEBS Lett. 144, 183-189.

WILSON, D.F. and FORMAN, N.G. (1982) Biochemistry 21, 1438-1444.

ZORATTI, M., PIETROBON, D. and AZZONE, G.F. (1982) Eur. J. Biochem. 126, 443-451.

Natural Oscillating Fields of Cells

HERBERT A. POHL
Oklahoma State University, Stillwater, OK 74078, USA

Some 14 years ago, H. Fröhlich (1968, 1980) predicted that living cells would be able to channel their chemical energies into high frequency electrical and cooperative oscillations. There is now increasing evidence that such electrical oscillations do occur. In particular, Fröhlich predicted that there would be a very high frequency oscillation in the frequency range of 100 gigahertz, but left open the possibility that electrical oscillations of a large scale and of a cooperative nature at yet lower frequency ranges could be observed. The evidence for the very high frequency range oscillations has been admirably summarized by Fröhlich (1980). More recent evidence from laser Raman spectroscopy (Webb, 1980) indicates that living cells, when metabolically active, do indeed produce oscillations at the very high frequency range. Del Guidice, Doglia, & Milani (1982) have suggested that the Fröhlich mechanism is augmented by solitons (Davydov, 1980) to assist in the production of low frequency oscillations.

Evidence for natural electrical cellular oscillations of a rather lower frequency can be inferred from several sources. Rowlands et alia (1982, 1982b, 1983) have demonstrated that long range interactions exist between live erythrocytes in their own fluid plasma. The measurements imply that the interactions range from 0 to 4 micrometers, a fact which would also strongly suggest the presence of long range cooperative interactions of the type proposed by Fröhlich.

Evidence for the presence of natural electrical oscillations at frequencies much lower than 100 gigaherz by cells has been obtained by research showing the presence of oscillating fields in and about cells. There are two phenomena which show such evidence; dielectrophoresis (DEP), and cellular spin resonance (CSR). Let us briefly summarize these lines of evidence.

Dielectrophoresis, the motion of neutral particles induced by the action of nonuniform electric fields (Pohl, 1951, 1978) can be used to explore for the existence of non-uniform electric fields. Here, the motion of tiny polarizable particles near cells is used to help decide if ac fields are present near the cell surface. The motion of tiny polarizable particles near cells is used to help decide if ac fields are pre-

Coherent Excitations in Biological Systems
Ed. by H. Fröhlich and F. Kremer

sent near the cell surface. The motion of such tiny polarizable neutral particles will be towards the region of higher field strength provided that their effective dielectric constant exceeds that of the suspending medium at that frequency. On the other hand, particles having an effective dielectric constant lesser than that of the suspending medium can be expected to experience a (mild) repulsion from the region of higher field intensity. Because this phenomenon is occurring on a microscopic scale, we will refer to it provisionally as "micro-dielectrophoresis" (micro-DEP). Experimental studies of micro-DEP on cells confirms the presence of natural rf oscillations in a number of ways. A preference for the accumulation of tiny (ca. 2 micrometer diameter) particles of high dielectric constant over that for ones of low dielectric constant:

1. Is shown by living but not by dead cells (Pohl, 1981b).
2. Can be suppressed in living cells by metabolic inhibitors such as those which block ATPases (Pohl et al. 1981).
3. Is maximal at or near mitosis, as demonstrated in the yeast (Saccharomyces cerevisiae) where the stages in the life cycle are readily visible by observing the bud development (Pohl et al., 1981).
4. Is suppressable, in accord with the theory of dielectrics, for a variety of test powders by increasing the effective dielectric constant of the medium through increasing the conductivity so that the effective dielectric constant exceeds that of the test powder. (Pohl, 1980a, 1980b, 1980c; Pohl et al., 1981). The effective dielectric constant is taken to be the absolute value of the complex dielectric constant, viz. $K = K'-jK''=K'-js/(\varepsilon_o\omega)$. $K_{eff} = <(k')+(k'')>^{\frac{1}{2}}$ where k is the complex relative dielectric constant, K" is the out-of-phase relative dielectric constant, j is the square root of minus one, s is the conductivity, ε_o is the absolute permittivity of free space, and ω is the angular frequency of the electric field.
5. Is observable for a variety of test powders of highly varied chemical nature, but that have the common feature of having a higher effective dielectric constant than that of the suspending medium. Such observations minimize the probability that the observed micro-DEP effects are due to chemically-related selective effects, and support the interpretation that the observed effects are due to dielectric factors, not chemical ones. For example, the highly polarizable particles which show preference for accumulation onto the living cells include $BaTiO_3$, $SrTiO_3$, $NaNbO_3$, and DP-1A, an organic polymer (Pohl and Wyhof, 1972) synthesized from anthraquinone and pyromellitic dianhydride, typical of class of aromatic polymers (Pohl and Pollak, 1977) having very high dielectric constants (1000 to 300,000). Among the types of particles of low polarizability, and which are but little attracted to living cells, and are used as "control" particles in the micro-DEP experiments are $BaSO_4$, silica, and corundum (alumina). (Pohl, 1980a, 1980b, 1981b; Pohl et al., 1981, Roy et al., 1981).
6. Is observable in a wide range of organisms, including bacteria, fungi, algae cells, and also avian blood cells and mammalian fibroblasts of normal, fetal, and tumor origin. (Pohl, 1980b, 1980c, 1981b; Pohl et al., 1981).

7. Provides micro-DEP versus conductivity data that yield estimates of the (minimum) frequency of the natural rf dipoles (in the neighborhood of 10 Khz) which agree with that deduced from results of CSR (Pohl, 1980c, 1981a).

8. Provides an estimate of the (minimum) field strength of the natural rf dipole field near the surface of the cells. This is derived from the known hysteresis of the field-dependent ferroelectric behavior of $BaTiO_3$. This value of the field at or near the cellular surface (about 100V/cm) agrees well with that inferred from torque measurements in CSR (Pohl, 1980c, 1981a).

It is worth pointing out that one expects the effect of micro-DEP to be of very short range, and to vary as the inverse seventh power of the distance between cell and the detector particle. It will be recalled that the DEP force varies as the gradient of the square of the field. Since the field about a dipole varies approximately as the inverse cube of the distance from the dipole center, the gradient of its square will vary as the inverse seventh power of the distance from its center. This implies that the DEP force upon the test particle (such as $BaTiO_3$, or SiO_2) can be expected to exert itself only over a very short range about the cells.

That the observed micro-DEP phenomena, the preference of highly polarizable particles of living cells over that for ones of low polarizability, is in fact an ac and not a dc phenomenon due to static fields is clearly evidenced by several facts. First, the observations cited in (4) above show that the presence of even slight conduction as by circa 100 micromolar NaCl or KCl is sufficient to suppress the effect, and is in accord with dielectric theory for ac fields. Second, it is well known from the theory of polar media such as water that the effect of static fields is readily suppressed over distances in the order of a Debye relaxation length. Since this is in the order of only about one hundred Ångstroms under our conditions, and the effects are seen to be active over distances which are many times larger (several micrometers), it is very unlikely that static effects are at issue. One may conclude that the evidence for natural rf fields in and about cells is supported by the evidence from micro-dielectrophoresis.

Let us now consider the evidence for natural electrical rf oscillations as obtained from the study of cellular spin resonance (CSR). The rotation, or spinning of cells or other particles in an ac field can provide a simple and straightforward method for obtaining the dielectric properties of individual cells (or particles). This is particularly true if rotating fields such as those produced by three or four-pole electrode arrangements are used. (Mischel, Voss, and Pohl, 1982; Arnold and Zimmerman, 1982a, 1982b,; Pohl, 1983).

Cells and other polarizable particles have been observed to spin while in a simple two-pole alternating electric field, and by a number of researchers. (Teixera-Pinto et al., 1960; Füredi and Valentine, 1962; Pohl and Crane, 1971; Pohl, 1978; Mischel and Lamprecht, 1980; Pohl, 1981a; Zimmermann et al., 1981; Pohl and Braden, 1982). Pohl and Crane first described the resonant spinning of live (yeast) cells that were subject to ac electric fields created between parallel wire electrodes, and noted that the spinning of the cells was sharply responsive to the frequency of the applied field. It was observed, moreover, that the resonant frequency depended upon the age of the colony examined. The sharply resonant character of the spinning response led to the use of the descriptive term, "cellular spin resonance" (CSR). Mischel and Lamprecht (1980) studied dielectrophoretic rotation rates of budding yeast (Saccharomyces cerevisiae) cells, but did not discuss the resonant rotation of the cells. Zimmermann, Vienken, and Pilwat (1981) reported CSR in several species of cells, including that of Friend cells, and human erythrocytes, and ghosts thereof; and mesophyll protoplasts of Avena sativa.

During the course of the experimental studies in the several laboratories, it was observed that there appear to be two circumstances for the occurence of CSR; that for lone cells little affected by local field perturbations due to delayed polarizations, and that for interacting particles for which delayed polarizations play a role in inducing the rotations. Physically, the latter case is rather analogous to the operation of a shaded-pole motor.

The spinning of lone cells was reported by Pohl and Crane (1971), by Chen, (1973), by Pohl and Braden (1982), and by Mischel and Lamprecht (1983). For physical reasons, only occasional and brief observations can be made of lone cells spinning in the mid region of a two-pole field, for the cells usually move quickly to an electrode or unite with other cells to form "pearl-chains". This is because of dielectrophoretic force or mechanical forces such as thermal upsets, ion injection, and the like involving streaming, etc. Lone cells can, however, be readily seen to spin freely and in a frequency resonant manner when against a mirror-smooth Pt electrode, as all investigators, and especially Mischel and Lamprecht (1983) have noted. The theory for the CSR of lone cells in a two-pole alternating electric field has been discussed by Pohl (1981a, 1982). That for the case of interacting particles exerting delayed polarization fields upon each other has been discussed by Holzapfel et al. (1982).

From the theoretical point of view, it is evident that cells can be made to spin by static fields or even very low frequency fields through the phenomenon of "rotational conduction". Here, the spinning of even spherical bodies is induced by action of ionic currents in producing surface-charge dipoles on particles having a conductivity or dielectric constant different from that of the supporting fluid medium. (Pohl, 1978, pp. 135-141). In this case it is expected that the spinning will be of low frequency. It is equally clear that CSR can also be evoked even in very high frequency fields if natural oscillating dipoles operating within living cells interact with the applied fields. The cellular spin rate would then be expected to be in the order of the difference between the frequencies of the natural dipolar frequency and that of the applied field. (Pohl, 1981, 1982). It is also clear that CSR can be evoked even in high frequency fields by delayed polarizations acting off-axis from that of the main ac field, in the manner discussed by Holzapfel et al. (1982). At the present writing, the only apparent mechanisms clearly accounting for the CSR of lone live cells where the spinning is sharply resonant to the applied frequency; where the spinning ceases upon the death of the cell, and where the frequency of cellular spinning is much less than the frequency of the applied field, would be ones in which a natural cellular dipolar oscillation exists and can interact with the applied field.

There exists, of course, the possibility that the natural cellular electrical oscillations might not be simply dipolar, but may be like that of linear quadupoles or of higher multipoles. It would also be expected that in view of the possibly weak character of the natural electrical oscillations that they would be affectable by applied fields and have their frequency "pulled" into resonance with applied fields, and so make the CSR spectrum consist of peaks somewhat broadened by the "pulling" or, as it is sometimes called, "frequency motor-boating".

The evidence that living cells naturally produce rf oscillations, as evidenced by CSR studies is supportive of the conclusions reached from the micro-DEP studies. On the other hand, the CSR evidence is not yet to be considered as strong as that from the micro-DEP, for there are several ways to interpret the CSR results. The observed CSR signals can be due, as discussed above to either the (lone cell) natural electrical oscillations or to the multicell delayed polarization interactions. In the latter case, no internal natural oscillations are required, but only the presence of the normal excess polarizability of the particles. The case is not yet closed, but the evidence so far from several laboratories does indeed show that lone cells, far removed from any visible perturbations which might be judged to be able to impart delayed polarizations, can spin resonantly with applied two-pole fields. (Pohl and Crane, 1971; Chen, 1973; Pohl and Braden, 1982; Mischel and Lamprecht, 1983). We conclude that the present evidence from CSR favors the idea that cells do indeed exhibit natural cellular electrical oscillations. Having said that, it must also be said that indeed most of the CSR experimentally observable in suspensions of cells, alive or dead, or of vesicles, or of cell ghosts, or even of inanimate particles is that due to the induced delayed off-axis polarization interactions, a rather easily evokable type of spinning. One must exercise considerable caution before assigning observed CSR as being due to the experimentally rarer and more difficult to determine natural rf oscillations.

One further remark on this point should be made at this time. Since it is possible that some of the natural rf oscillations might be quadrupolar as well as dipolar in character, it may eventually be found that special conditions are necessary to observe them in some cases. The situation could be considered somewhat analogous to the IR and Raman spectra of molecules. The dipolar state can be observed by direct emission, but the quadrupolar states may need to be stimulated to be observable. Further study along this line is necessary.

On using two-electrode systems to examine the CSR of living cells, Pohl and Braden (1982) observed in the case of yeast (Saccharomyces cerevisiae) cells that the spectrum typically contained only one or two peaks. The cells spun at a rate of about 0.1 to 30 Hz while the applied frequency was in the order of thousands of Hertz. Moreover the spin response was sharply resonant to the applied frequency. The CSR was observed to vary with the cell type, and with the phase of the yeast cell life cycle. Living cells responded readily. Dead cells showed little or no such response when observed as lone cells, well removed from others. Live mouse sarcoma cells also showed sharp CSR. As an additional test of the presence of natural rf oscillations, these researchers compared the CSR of cells in the presence of pulsed dc and of sinusoidal rf fields. They showed that the ac and pulsed dc spectra are quite similar, and that the cells continued to spin in either field of the same basal frequency and applied field strength (14 V rms). If cells were to possess induced-polarization dipoles only, they would be expected to librate but not spin as lone cells in the high frequency pulsed dc fields. Dead lone cells, killed as by heat or by phenol were observed not to spin when subject to high frequency electric fields, ac or pulsed dc. The natural electrical cellular oscillations can be considered to arise by a number of possible mechanisms. In considering any particular case, it is important to appreciate that the field effects upon external objects must be considered to be due to electrical distortions at or near the cellular surface, for the high conductivity of the cell interior can be expected to mask or strongly damp the fields arising from deep within the cell. Moreover, the field effects upon external objects such as the test particles used in micro-DEP can also be expected to be of very short range unless the conductivity of the support medium is very low. Said another way, one can expect that only surface-connected asymmetric charge distributions would be sensed by nearby bodies; and that the external effects of natural cellular rf oscillations will be short-ranged and highly ineffective unless in aqueous media of very high resistivity. The range of natural rf electrical oscillations will be longer the higher the frequency. Among the mechanisms which can be considered to give rise to natural rf oscillations are those involving the Fröhlich mechanism of cooperative dipolar oscillations fed by cellular metabolic energy, and mechanisms involving oscillating chemical reactions capable of producing charge waves. The latter may well involve limit cycle operation as discussed by Fröhlich (1977) and by Kaiser (1982). A model for oscillating chemical reactions occuring in a cell and giving rise to charge waves was discussed by Pohl and Braden (1982).

In the latter model, it was suggested that natural intrinsic oscillating cellular dipoles arise from oscillating chemical reactions coupling within the cells to physically mobile regions of ions so as to produce charge density waves. There are several well-studied systems of oscillating chemical reactions (Noyes and Field, 1974; Noyes, 1977; Epstein et alia, 1981). These well-studied oscillating reactions are considered to oscillate between phases which are alternately rich in free radical and ionic species (Fig. 1). Such reactions have both temporal and spatial coordinates, and develop in both space and time. It is suggested that if, during the ionic phase of the reaction cycle, the outward speed of the positive ions does not exactly match that of the negative ions, then a charge wave will develop (Fig. 2).

Fig.1.

Diagram of reaction sequences:
Curve A. The usual fall of reactant concentration with time.
Curve B. The fall of a reactant concentration during a periodic reaction in which no externally supplied reagents are added during the reaction.
Curve C. Free radical branching reaction phase

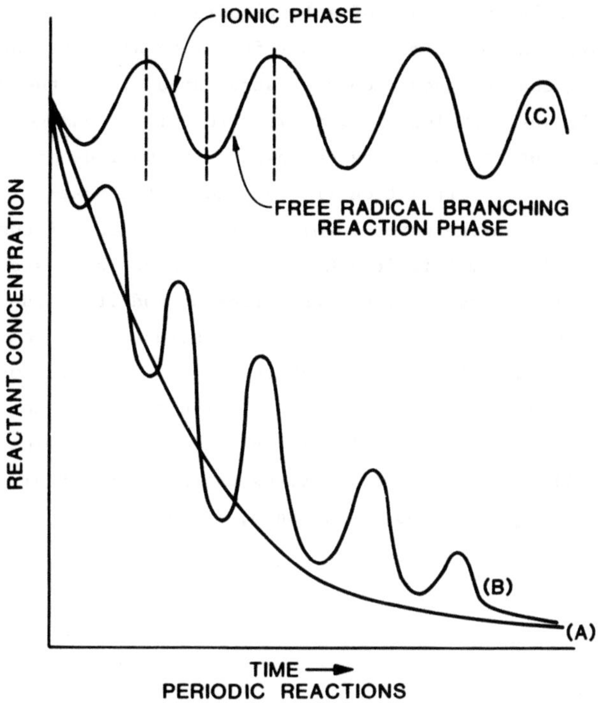

Fig.2.

A mechanism for the production of natural oscillating electrical dipole oscillations in cells or organelles.
A. Charge waves from the ionic phase of periodic reaction radiate outward from initiating sites within the cell. Initially the charge waves are uncollimated and incoherent.
B. The presence of the various more or less parallel structures within the cell causes the individual charge waves to become parallelized. Whether the reaction waves proceed parallel or at right angles to the retaining walls of the structures is not yet known, but present evidence from kinetic considerations would indicate that the reaction proceeding at right angles to the restricting wall is the more likely. For ease of picturing the process of the collimation and coherence, however, the parallel course of the reactions is sketched.
C. Development of the collimation and coherence of the individual charge waves is pictured.

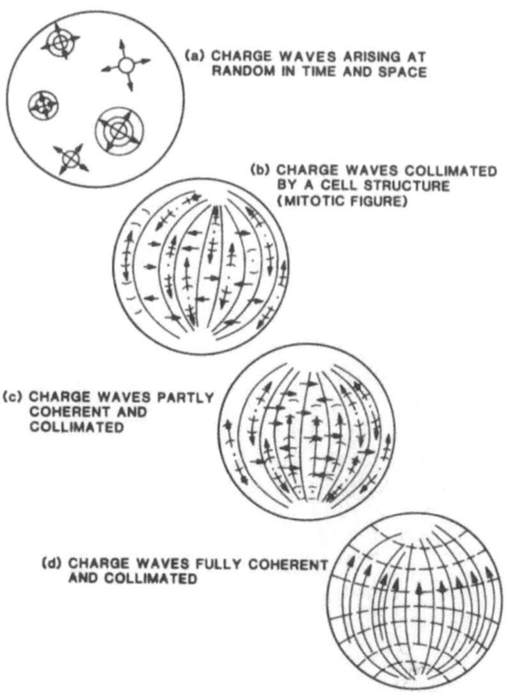

D. Full coherence and collimation of the charge waves. It is expected that the resultant dipole (or higher multipole) oscillator will be detectable outside the cell only from charge wave effects very close to the cellular surface, for the conductive nature of the cell plasma and normal surroundings will be expected to strongly damp low frequency oscillations.

This is an example of a longitudinal electric wave. Such charge waves can, of course, arise almost simultaneously from several sites throughout the cell. As these developing charge waves encounter structures within the cell that effect a parallelization of the wave developments, collimation of the charge waves will ensue. Because of coulombic interactions, one can expect the various collimated charge waves to develop a cooperative correlation (in the sense of a Bose-like condensation) to form a coherent set of charge waves involving large regions of the cell in a cooperative oscillating dipole or multipole. This testable model suggests a number of experimental studies as to its energy source, its strength, and its cause.

There are a number of well known structures in cells which might be able to effect parallelization of such charge waves. These include the mitotic spindle apparatus, the walls of the endoplasmic reticulae (ER), the laminae of the cristae in mitochondria, and the grana of chloroplasts (Fig. 3). Oscillating reactions in biology are well known. Excellent reviews on this subject are available. (Treherne et al., 1979; Berridge and Rapp, 1979; Rapp, 1979).

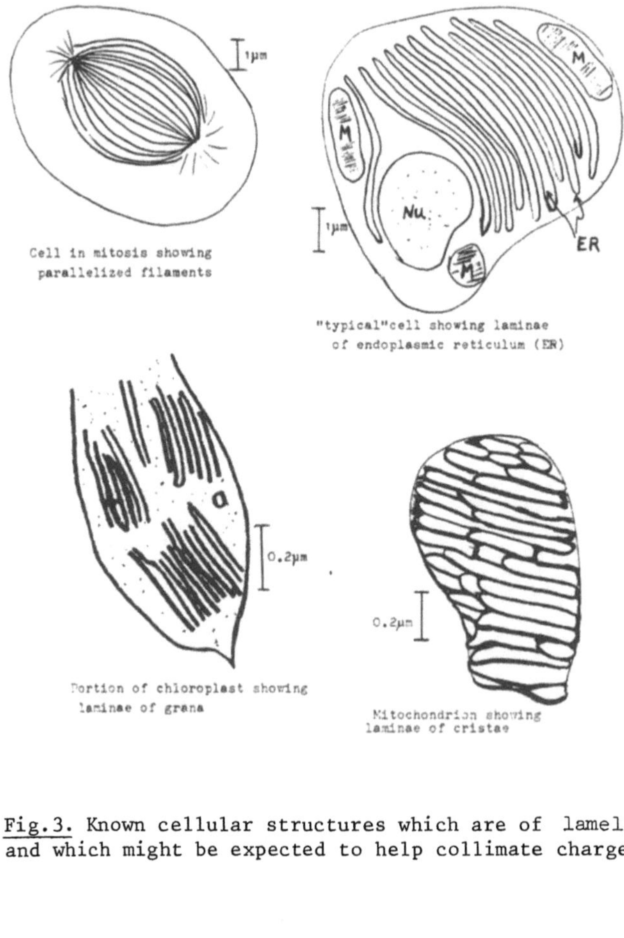

Fig.3. Known cellular structures which are of lamellar or parallelized morphology, and which might be expected to help collimate charge waves and periodic reactions

Whether the proposed charge waves providing the source of the natural rf oscillations are due to reactions which proceed geometrically parallel to or at right angles to the laminar structures is not known, but kinetic data on the assumed charge waves imply that the charge waves proceed across rather than along the laminae, especially in the case of the high frequency oscillations. A study by Schmidt and Ortoleva (1979) suggests how oscillating reactions might be coupled to electrical charge waves. It is probable that the ubiquitous ionic double layers at the numerous interfaces within the cells would play a role in the natural oscillations of cells. In this connection, it may be noted that the mechanically induced electromagnetic radiation from vibrated systems having ionic double layers may play a role in cellular responses (Pohl, 1980a). Another mechanism for producing resonant oscillations is that which could arise from local fluctuations in the charge density at the surface of the cell, which in turn arise from internal reaction oscillations. Consider a local charge density fluctuation dq, which arises at some point at the cell surface at time t=0. It will affect the charge distribution around the periphery of the outer ionic double layer of the cell-medium interface. This distribution will travel around the cell surface and reverberate back to the origin with a frequency f, in the order of $f=v/\ell$, where v is the velocity of the disturbance, and ℓ is the half-perimeter. A rough estimate of the frequency of this reverberation can be made by assuming that the velocity is in the order of that of sound, and the distance ℓ, is of the order of π times the radius. Then for a cell of radius 10 microns, the frequency of such a reverberation will be in the order of $f=v/\ell \cong (100\ m/s)/(3.14 \times 10\ microns) \cong 30$ MHz (Fig. 4).

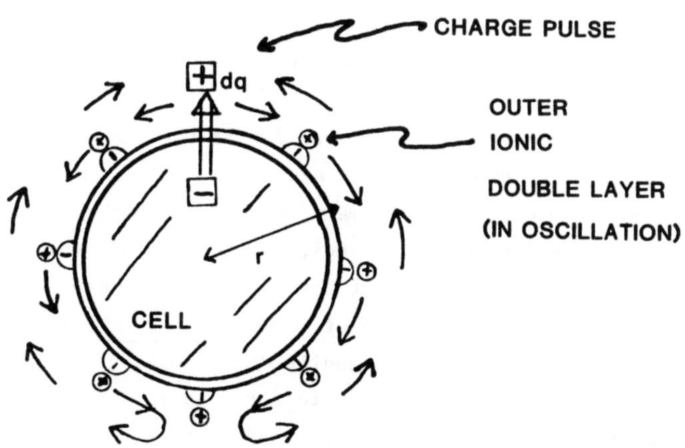

Fig.4. Diagram of a mechanism for producing pulses of oscillatory character that originate from a charge ejection (dq) from some site. The appearance of the surplus charge can then evoke a charge wave reverberating around the periphery in the ionic double layer at the cell-medium boundary

More realistic values of the velocity of propagation of the disturbance along the ionic double layer would be expected to lead to yet lower values of the reverberation frequency of this type of electrical oscillation. The intensity of this mode will depend upon the feedback between the internal and external waves.

We have before us, then, the real probability that cells produce oscillating electric fields. It is appropriate to ask where, how, and why such oscillations might arise in living systems. There appear to be two main experimental results available to us at this point in time. The first broad result from the data to date is that living cells of many types give evidence of the presence of natural oscillating electric fields. From studies of micro-dielectrophoresis, it is observed that the particle-gathering ability of various cells is one which favors the accumulation of the highly polarizable particles over that of the weakly polarizable particles. This is shown in Table 1.

TABLE 1.

The Particle Gathering Ability of Various Cells
...Micro-dielectrophoresis

CELL TYPE	RATIO, POLARIZABLE TO NON-POLARIZABLE PARTICLES; (n/p) polar to (n/p) non-polar.
Bacteria, *Bacillus cereus*, synchronized	4
Yeast, *Saccharomyces cerevisiae*	2.0
Alga, *Chlorella pyrenoidosa*	2.2
Avian, Chicken red blood cell	2.0
Mammalian, Mouse "L" fibroblasts	2.0
Mammalian, Mouse Ascites tumor fibroblasts	2.2
Mammalian, Mouse, Fetal fibroblasts	2.0

From the above table it is observed that the process is ubiquitous, and common to all of the wide types of cells examined.

A second line of evidence comes from micro-dielectrophoresis observations (Pohl et alia, 1981) where the n/p values, indicating the strength of preference of cells for polar particles, were found to be maximal at or near mitosis. These results were readily obtained in the case of yeast, *Saccharomyces cerevisiae*, in which the phase of the life cycle is readily determinable from the shape of the cell. The life cycle variations of the micro-dielectrophoresis for $BaTiO_3$ clearly peaked at the period of the life cycle when the small budding stage is reached, i.e., when mitosis is nearby in the life cycle.

With these two principle results before us, i.e., (1) that natural cellular electrical oscillations are "universal" among cells, and (2) that the oscillations are maximal at or near mitosis; we may ask the question: "Are the observed natural rf oscillations evidence of a process which is a necessary one, or is it evidence of just an unnecessary frill in the life processes of cells?" The fact that they are seen in such a wide variety of cells, ranging from the primitive bacteria to the sophisticated mammalian cells, indicates that the natural rf oscillations have been carried on for billions of years, and that they could reflect something or some process essential to cell life. The second fact, that such natural rf oscillations are maximal at or near mitosis can be taken to imply that the oscillations reflect a process important to cellular reproduction. It remains then, to find why and how these natural cellular rf oscillations play an essential role in cellular replication. As we learn more about the answers to these questions, we shall understand more about factors governing the replication of cells during the four critical phases of higher life forms: during embryonic development, during somatic repair, during wound healing, and during tumor growth.

Helpful discussions were had with Professor H. Fröhlich and I. Lamprecht, Dr. F. Kaiser, and V. Denner. Support of a portion of this research by the Pohl Cancer Research Laboratory is gratefully acknowledged. The newly reported measurements of micro-DEP on algae, and yeast were made by Hiram Rivera of the Laboratory.

BIBLIOGRAPHY

Arnold, W. M., and Zimmermann, U. (1982a) Naturwissenschaften 69, 297.
Arnold, W. M., and Zimmermann, U. (1982b) Z. Naturforsch. 37c, 908.
Berridge, M. J. and Rapp., P. E. (1979) J. Exp. Biol. 81, 217.
Chen, C. S. (1973) "On the Nature and Origins of Biological Dielectrophoresis", Ph.D. Thesis, Oklahoma State University, Stillwater, OK 74078.
Davydov, A. S. (1980), Sov. Phys. JETP, 51, 397.
Del Guidice, E., Doglia, S., and Milani, M. (1982) Physica Scripta 26, 232.
Epstein, I. R., Kustin, K., de Kepper, and Orban, M. (1981) J. Amer. Chem. Soc. 103, 2133.
Fröhlich, H., (1968) Int. J. Quantum Chem. 2, 641.
Fröhlich, H., (1977) Neurosci. Res. Program Bull. 15, 67.
Fröhlich, H., (1980) Adv. Electron. Electron Physics 53, 85.
Füredi, A. A. and Valentine, R. C. (1962) Biochem. Biophys. Acta 56, 33.
Holzapfel, C., Vienken, J. and Zimmermann, U. (1982) J. Membrane Biol. 67, 13.
Kaiser, F. (1982)(this symposium 1982, this Journal, 1983).
Mischel, M. and Lamprecht, I. (1980) Z. Naturforsch. 35c, 1111.
Mischel, M., Voss, A., and Pohl, H., (1982) J. Biol. Physics 10, 223.
Mischel, M. and Lamprecht, I. (1983) "Rotation of cells in nonuniform rotating alternating fields", J. Biol. Physics, 11 (in press).
Mischel, M. and Pohl, H. (1983) "Cellular spin resonance in rotating fields", J. Biol. Physics, 11. (in press).
Noyes, R. M. and Field, R. J. (1974) Ann. Rev. Phys. Chem. 25, 95.
Noyes, R. M. (1977) Acc. Chem. Res. 10, 214, and 273.
Pohl, H. (1951) J. Appl. Phys. 22, 869.
Pohl, H. and Crane, J. S. (1971) Biophys. J. 11, 711.
Pohl, H. and Wyhof, (1972) J. Non-Cryst. Solids, 11, 137.
Pohl, H. and Pollak, M. (1977) J. Chem. Phys. 66, 4031.
Pohl, H. (1978) "DIELECTROPHORESIS, The Behavior of Matter in Nonuniform Electric Fields", Cambridge University Press.

Pohl, H. (1980a) "Micro-dielectrophoresis of dividing cells", in BIOELECTROCHEMISTRY, pp. 273-95, edited by H. Keyzer and F. Gutmann, Plenum Press.
Pohl, H. (1980b), Int. J. Quantum Chem. 7, 411.
Pohl, H. (1981a), J. Theor. Biol. 93, 207.
Pohl, H. (1981b), J. Bioenerg. Biomembranes 13, 149.
Pohl, H. Braden, T., Robinson, S., Piclardi, J., Pohl, D. G., (1981), J. Biol. Phys. 9, 133.
Pohl, H. (1982) Int. J. Quantum Chem. 9, 399.
Pohl, H. and Braden, T. (1982) J. Biol. Phys. 10, 17.
Pohl, H. (1983) "Cellular spinning in pulsed rotating electric fields", J. Biol. Physics, 11. (in press).
Rapp., P. E. (1979), J. Exp. Biol. 81, 281.
Rowlands, S., Sewchand, L. S., Lovlin, R. E., Beck, J. S., and Enns, E. G., (1981) Phys. Lett. 82A, 436.
Rowlands, S., Sewchand, L. S., Enns, E. G. (1982) Can. J. Physiol. Pharm. 60, 52.
Rowlands, S., (1982) J. Biol. Phys. 10, 199.
Rowlands, S., Eisenberg, C. P., and Sewchand, L. S. (1983), "Contractils: Quantum mechanical fibrils", J. Biol. Phys. 11. (in press).
Roy, S. C., Braden, T., and Pohl, H., (1981) Phys. Lett. 83A, 142.
Schmidt, S. and Ortoleva, P. (1979), J. Chem. Phys. 71, 1010.
Teixera-Pinta, A. A., Nejelski, L. L., Cutler, J. L., and Heller, J. H. (1960), Exp. Cell Res. 20, 548.
Treherne, J. E., Foster, W. A. and Schofield, P. K. (1979), "Cellular Oscillators" J. Exp. Biol. 81 (review volume).
Webb, S. J. (1980) Phys. Rep. 60, 201.
Zimmermann, U., Vienken, J. and Pilwat, G. (1981), Z. Naturforsch. 36c, 173.

The Interpretation and Use of the Rotation of Biological Cells

U. ZIMMERMANN and W. M. ARNOLD

Arbeitsgruppe Membranforschung am Institut für Medizin, Kernforschungsanlage Jülich GmbH, Postfach 1913, D-5170 Jülich

Introduction

The application of electric fields to cell suspensions has resulted in the evolution of three new techniques - those of cell rotation, reversible membrane breakdown and fusion. Although these methods differ in the order of magnitude of field involved, they have several concepts and physical parameters in common. Owing to the subject area of this conference, neither breakdown nor fusion will be discussed here, and have already been reviewed earlier (1, 2, 3, 4).

Rotation can be used as a tool to provide experimental membrane data useful for fusion or other purposes, if the theory underlying it is understood. The experimental grounds on which this theory rests will therefore be discussed first.

Experimental Basis

Teixeira-Pinto et al. (5) were the first to report that living cells may rotate when exposed to a linear alternating field. This rotation was observed only after the field either had caused two cells to approach each other very closely or had first detached a fragment from an Amoeba. Although polystyrene spheres (and many other materials in suspension, see e.g. (6)) show this "pearl-chaining" effect, rotation was apparently restricted to cells or vesicles formed from cells. Teixeira-Pinto et al. used the 1-100 MHz frequency range and parallel, glass-covered electrodes (Fig. 1a).

Coherent Excitations in Biological Systems
Ed. by H. Fröhlich and F. Kremer

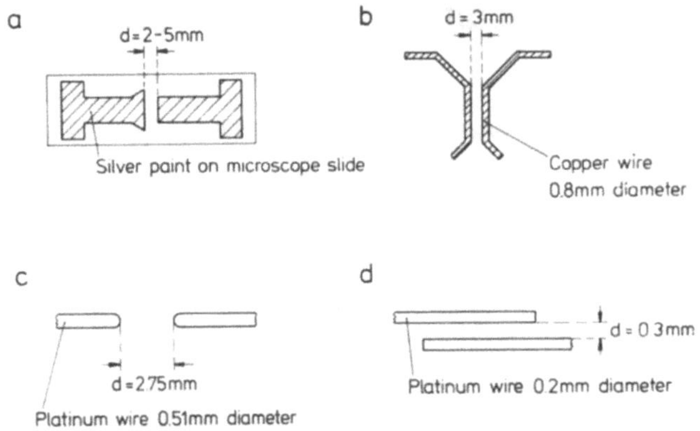

Fig. 1. Electrode arrangements which have been used to apply fields resulting in cell rotation. The parallel arrangements in (a), (b) and (d) will result in fields which are relatively uniform compared to that produced by the pin-pin electrodes in (c). (a), (b), (c) and (d) are based on (5), (6), (8) and (12) respectively

Füredi and Ohad (7), using similar electrodes (Fig. 1b), observed a sort of rotation or realignment of erythrocytes by a mechanism which appeared to involve deformation of the cell. As it is stated that application of the 120 MHz field for longer than 5 seconds caused break-up of the cells, it appears that the voltage levels involved were causing artifacts due to heating or membrane breakdown (1). Whilst working with highly non-uniform fields (designed to investigate dielectrophoresis) produced between two pin electrodes (Fig. 1c) Pohl and coworkers (8, 9, see also 10) found that suspended yeast cells could be made to rotate anywhere in the field between the electrodes. At any given frequency (which could be in one of two or three conductivity dependent regions between 100 Hz and 100 kHz), a small proportion of the cells rotated.

Mischel and Lamprecht (11) confined their observations to single cells on the surfaces of pin electrodes. They reported that those cells seen to rotate did so with a speed linearly proportional to field strength, once a threshold was exceeded. The observation that the magnitude of the threshold was affected by the moment of inertia (which by definition affects only the acceleration of each cell) suggests that the spinning of the cells was not continuous. Presumably the strong attractive force exerted on the cells by the electrode, combined with the irregular shape of the "budding" yeast cells, caused intermittent hindrance to rotation. Indeed the authors noted that rotating cells might suddenly stick to the electrode.

Predictable, reproducible rotation of all cells in a given population was observed by Zimmermann et al. (12) upon application of linear fields to high suspension densities of yeast cells, erythrocytes, plant protoplasts or Friend cells. (A permanent cell line consisting of mouse erythrocytes made leukaemic with Friend virus). At a given conductivity, each cell type seemed to have its own particular optimum ("resonant") frequency. The formation of pearl chains of cells was seen to hasten the onset of rotation, as if rotation were a collective phenomenon. Hub et al. (13) have also observed the rotation of liposomes.

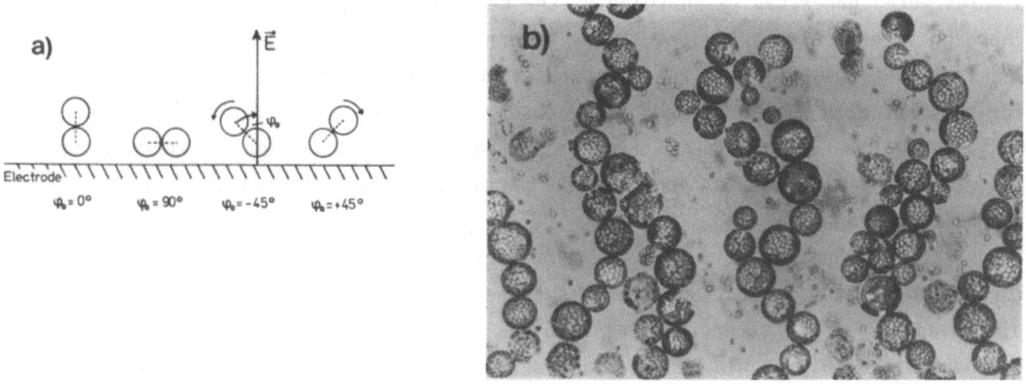

Fig. 2. Theoretical and observed orientation of cells in an electric field in association with multi-cell rotation. In (a), the angle φ_o defines the orientation of a doublet of cells with respect to the electric field vector. According to (14), positive values of φ_o give clockwise rotation of both cells, negative values give anti-clockwise rotation. Maximum rotation speed occurs at $\pm$ 45°. In (b) the orientation of plant protoplasts taken up during multicell rotation is shown, the electrodes being horizontal. Fig. taken from (14)

Subsequently, Holzapfel et al. (14) investigated whether the rotation could be due to the induction of dipoles in the cells by the field. The theory developed showed that a torque should indeed be developed provided that a minimum of two cells were in close proximity, and that the torque acting on the cells of any doublet should be maximized when the axis of the doublet was at an angle of 45° to the field (Fig. 2a), as borne out by observation (Fig. 2b). The direction of rotation is determined by whether the axis is displaced one way or the other from the field. In addition, the frequency maximum for cells of a given type was shown to be due to the production of maximum torque by the dipole-dipole interaction when the field frequency (f) matched the relaxation time (τ) of the polarisation process of the cell. Specifically maximum rotation rate should occur when

$$2\pi f\tau = 1 \quad (1)$$

It is possible to view the dipole-dipole-field interaction as producing a local rotating component in the electric field. According to this, any single member of a given cell type should rotate if a rotating field (pulsed or continuous) of the appropriate frequency is applied, as was recently demonstrated in this laboratory (continuous field method (15,16); pulsed field method (17)). A single cell or an entire cluster of cells were seen to rotate in a direction established immediately by the external field. Using the continuous rotating field method (4, 15, 16), it was also shown that the field frequency giving fastest rotation was a linear function of medium conductivity as was to be expected (see next section) if the dipoles were generated by charging of the cell membrane, because the rate of charging is effectively limited by the external conductivity, when this is low. The view that possession of an intact outer membrane was the essential requirement for this rotation was confirmed by the observation that isolated plant cell vacuoles or large unilamellar artificial liposomes also rotate in the same frequency range of the rotating field, whereas ion-exchange particles or polystyrene beads do not (unpublished results).

The theories of multi- and single-cell rotation both predicted that the torque and hence the rotation rate of suspended cells should increase as the square of the applied field strength (4, 14, 15, 16). Observations on cells where only hydrodynamic damping could be operating showed this prediction to be correct, there being no detectable threshold (Figs. 3a, 3b). The linear results (with a substantial threshold) obtained by Mischel and Lamprecht (11) were presumably a consequence of the strong interference to rotation caused by the dielectrophoretic force exerted on cells when next to a pin electrode.

It is certain from the above that all cells can be made to rotate by the application of a rotating field, and that in so rotating they are behaving in a passive and linear manner. The occasional rotation of even single cells in a <u>linear</u> field as reported by Pohl and co-workers (8, 9) and by Mischel and Lamprecht (11) is more difficult to interpret. Pohl has put forward several hypotheses to account for such rotation (10, 18, 19) the most recent of which calls for the <u>intrinsic</u> production of radio frequency (5 kHz or higher) oscillations within the cell. Such oscillations are supposedly driven by corresponding oscillations of a biochemical nature, presumably coupled by means of charge displacement across internal membranes. However it is unfortunate that at present there seems to be no report of fast enough biochemical oscillations: the periods of this class of oscillation being at least several seconds (20). In addition, Pohl's proposed "motorboating" phenomenon (9, 18, 19) appears to contravene the normal phase relationships in the production of difference frequencies. Pohl and Braden (9) predicted that the frequency with which the cell rotates (f_R) should be determined by the difference between the frequencies of the applied field (f_E) and that of the cell's inherent oscillator (f_D):

$$f_E = f_D \pm f_R \quad (2)$$

If the rotation rate depended upon the difference frequency between the inherent and applied frequencies, the direction of rotation should reverse sharply as the applied frequency is swept across the inherent frequency. No such sharp reversal has been reported.

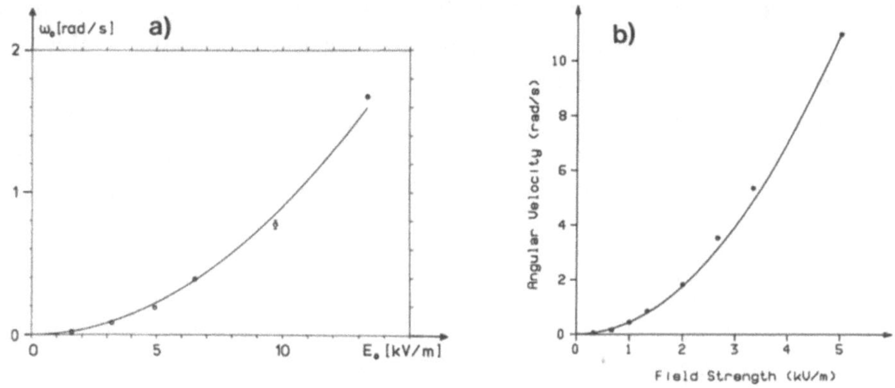

Fig. 3. The square-law dependence of rotation rate on field strength for mesophyll protoplasts of Avena sativa (approx. 18 µm radius).
a) Multi-cell rotation, linear field (redrawn from (14)).
b) Single cell rotation, rotating field (taken from (15)). The lower field strengths necessary for single-cell rotation are evident

It should be emphasized that any chamber (even one designed to produce a uniform field at the centre) is likely to have regions where the field intensity is much above the mean value. This can lead to localised heating resulting in turbulence throughout the chamber, although the effect is usually strongest close to the electrodes where membrane breakdown may cause electrolyte release giving yet further heating. Turbulence may also result from heating due to illumination without adequate heat filtration. We believe this may have given rise to some of the effects reported, and consider that only highly repeatable rotation, using any of the cells from a uniform population, should in future be reported.

A possible dielectrophoretic cause for the rotation of single cells near to pin-pin electrode systems may be found in the strong inhomogeneity of the field, seen especially at close electrode spacings. Lafon and Pohl (21) have shown that when cylindrical or pin electrodes induce image charges in each other, the effect is to add to the field expected as the common axis of the electrodes is approached. The side of a cell nearer to the electrode axis will therefore experience a greater force towards the locally dominant electrode than will the other side, in other words a torque is exerted on the cell. A cell at the surface of an electrode can therefore be expected to roll towards the central axis. Suspended cells close to the electrode should rotate as they move towards it, although crossing the field of view of a microscope too quickly to allow repeatable observations.

It may be seen from the brief review given above that the rotation of single cells in two electrode chambers has only been reported by users of the pin-pin electrode pattern (Fig. 1c). The field between such electrodes will usually be much more non-uniform than that between parallel electrodes (Figs. 1a, 1b or 1d). This deduction would seem to support the above non-uniform field explanation of these (8, 9, 11) observations, which may therefore be justly termed dielectrophoretic rotation.

We wish to conclude that using accepted field theory it is possible to explain all of the repeatable observations of rotation in linear fields, it being our impression that the remaining observations are not readily repeatable. However, the consistent theories that account for the highly repeatable phenomena of multi-cell rotation in an alternating field and of single cell rotation in a rotating field have become the basis of quantitative techniques, as described in the next section.

Theory and Application

The theory of multi-cell rotation developed by Holzapfel et al. (14) showed that the rotation speed was maximised at a field frequency which matched a generalised dipole relaxation time (τ). We shall term the frequency giving maximum rotation rate the characteristic frequency (f_c). The value of f_c is given by Eq.(1).

This same equation was later derived also for single cell rotation (4, 16), and the polarisation mechanism shown to be charge separation across the membrane. In multi-cell rotation, the effective τ value is expected to depend on all cells present and this prevented the quantitative application of this technique in heterogeneous populations. Moreover, in low conductivity (10^{-5} S$\cdot$ cm^{-1} or less) media the τ values for multi-cell rotation seem to be several times shorter than those expected from the single cell theory or found in single cell rotation experiments (below). This may correspond to the cells being more conductive than these media due to membrane resistance or ionic double-layer effects. Whatever the reason, it is important to note that the multi-cell rotation will give a value for τ which is appropriate to the usual conditions for cell fusion (several cells in contact in low conductivity media).

When a spherical membrane-bounded object (of radius a (metres), membrane capacitance and conductance C_M (F$\cdot$cm^{-2}) and R_M ($\Omega\cdot$cm ^{2}) respectively and of internal conductivity σ_i (S$\cdot$cm^{-1})) is placed in a medium of conductivity σ_e S$\cdot$cm^{-1}, the relaxation (charging) time of the membrane in response to electric fields in the medium is given (22) by:

$$\tau = a C_M \frac{\sigma_i + 2\sigma_e}{2\sigma_i\sigma_e + \frac{a}{R_M}(\sigma_i + 2\sigma_e)} \qquad (3)$$

The processes that this equation describes become clearer when it is viewed as an expansion of an equation developed by Jeltsch and Zimmermann (23):

$$\frac{1}{\tau} = \frac{1}{\tau_e} + \frac{1}{\tau_i} \qquad (4)$$

In this equation, τ_e is the time constant of the membrane (considered to be a lossless capacitance) in series with the internal and external solution resistances. τ_i is the correction that must be made because of the finite resistance of the membrane: it is conceptually and numerically equal to the $R_M C_M$ time constant. In the case of a maximum in rotation rate Eq. (3) can be combined with Eq. (1), giving:

$$f_c = \frac{1}{2\pi a C_M}\left[\frac{2\sigma_i \sigma_e}{\sigma_i + 2\sigma_e} + \frac{a}{R_M}\right] \tag{5}$$

If the external medium is made considerably less conductive than the cell interior, then

$$f_c = \frac{1}{2\pi a C_M}\left(2\sigma_e + \frac{a}{R_M}\right) \tag{6}$$

In the majority of cells, R_M is large enough to be neglected and so we obtain:

$$f_c = \frac{\sigma_e}{\pi a C_M} \tag{7}$$

Arnold and Zimmermann (16) were indeed able to demonstrate a linear relationship between the characteristic frequency and external conductivity for single-cell rotation (Fig. 4a), the value of membrane capacitance derived (Eq. (7)) being 0.48 ± 0.07 µF· cm^{-2} with data from other plants yielding a value of 0.37 ± 0.03 µF· cm^{-2} (unpublished results). The cells used, mesophyll protoplasts, have a vacuole which occupies at least 70% of the cell, so the capacitance of the individual (effectively series-connected) membranes can be estimated as approximately 1 µF·cm^{-2} each.

This constitutes the first method for the determination of the membrane capacitance of very small quantities of intact cells (i.e. without the use of intracellular electrodes). A value of 1.0 µF·cm^{-2} has also been obtained by work (Arnold, Küppers, Wendt and Zimmermann, unpublished) on Friend cells (Fig. 4b). These cells have no large internal membrane system, so the capacitance obtained is that of the plasmalemma.

Knowledge of the membrane capacitance is required to understand electrically induced membrane breakdown and cell fusion, because the capacitance determines the charging time for the membrane (Eq. (3)) and therefore the minimum length of breakdown pulse necessary.

Biomembrane capacitance is thought to be considerably affected by the protein content (24). If proteins within the membrane are considerably hydrated, or form aqueous pores, then their relative dielectric constant up to frequencies of at least several MHz should be much higher (25) than the value of 2 usually accepted for pure lipid bilayers. On the other hand, surface layers of protein should decrease the overall capacitance because at sufficiently high frequencies they will appear as simple capacitive elements in series with the membrane.

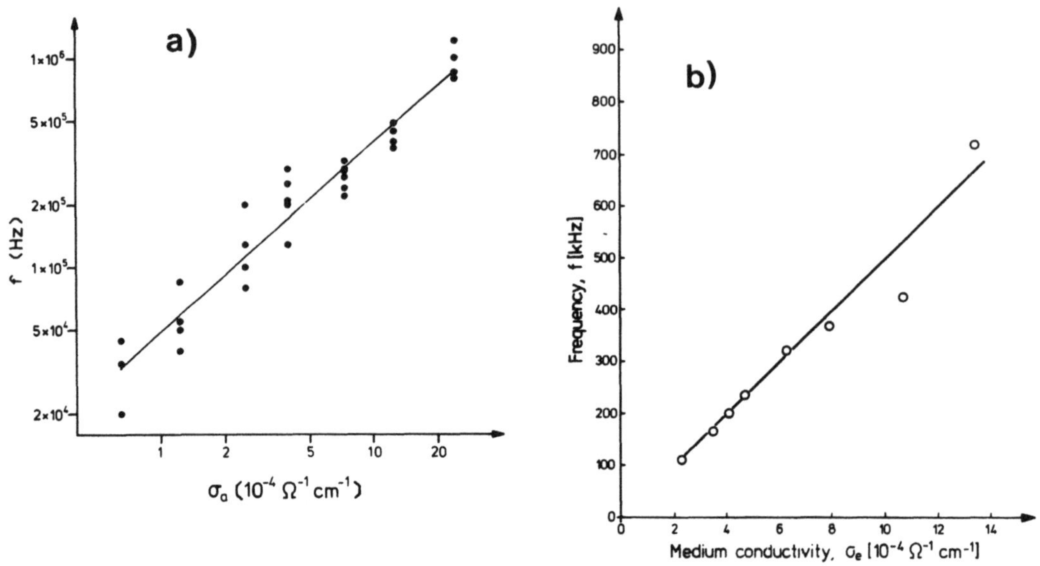

Fig. 4. The dependence of characteristic frequency on medium conductivity measured using single cells in the rotating field. a) Values for individual mesophyll protoplasts of A. sativa, axes are double logarithmic. The slope corresponds to a membrane capacitance of 0.48 ± 0.07 $\mu F \cdot cm^{-2}$ (from (14)). b) Values (each is the mean of 6 cells) for Friend cells. The slope corresponds to a membrane capacitance of 1.0 µF. Repetition of this experiment five times has shown that the deviations from the straight line are random and seem to result from scatter in the cell population itself. Mean and standard deviation of the membrane capacitance for the five runs, 0.99 ± 0.08 $\mu F \cdot cm^{-2}$

The presence of protein-free regions within the membrane is thought to be necessary for the process of electrofusion (2, 3, 4). Therefore it can be expected that capacitance measurements will promote the understanding of the electrofusion process by revealing something of the membrane protein distribution.

According to Eq. (6), the membrane resistance value may also be determined by cell rotation experiments. The effect of this resistance so far observed has been too small to yield an accurate value for R_M although a lower limit of 30 $\Omega\, cm^2$ (Arnold, Schnabl and Zimmermann, unpublished results) may be placed on it (Vicia faba mesophyll protoplasts). In other cases, when R_M is high the term in R_M in Eq. (5) may be omitted if the lowest conductivities are not to be considered. Then rearrangement yields:

$$\frac{1}{f_c} = \pi a C_M \left(\frac{1}{\sigma_e} + \frac{2}{\sigma_i}\right) \tag{8}$$

Hence σ_i may be determined, and values for C_M derived even if σ_i cannot be assumed to be much larger than σ_e. However, accurate determination of σ_i requires the use of such high medium conductivities that heating effects (turbulence and temperature changes) at present limit this.

In the special case where $\sigma_i = \sigma_e$ (= σ), Eq. (5) reduces to

$$f_c = \frac{1}{3\pi a C_M} (\sigma + \frac{3a}{2R_M}) \tag{9}$$

which should be applicable to cells which have equilibrated their contents with the medium, yet still have an electrically functioning membrane. An example of such should be resealed erythrocyte ghost cells, although so far the rotation of these has been attempted using the multi-cell technique only. Liposomes suspended in the medium in which they were prepared should also have equal internal and external conductivities. It should thus be possible to find the membrane resistance by forming ghosts or liposomes in media of a series of conductivities, and plotting the rotational characteristic frequency against the conductivity. In the limit when the R_M is highenough to be neglected:

$$f = \frac{\sigma}{3\pi a C_M} \tag{10}$$

which means that the time-constant for such liposomes is three times longer than that expected for cells of the same size (c.f. Eq. (7)). This is of interest when considering the conditions necessary to fuse such liposomes.

The equations above describe rotation having a frequency-dependence which is due to field-driven charging of the membrane through internal and external medium resistances. As discussed in (16), the frequency dependence of the torque on the cell is given by:

$$N = E^2 P \frac{2\pi f \tau}{1+(2\pi f \tau)^2} \tag{11}$$

where N is the torque, E is the field strength and P is a constant for each cell.

Eq. (11) has the same form as that of a Debye relaxation (involving field-dependent re-orientation of dipoles having a single relaxation time). The condition for maximum torque in the rotating field is that the field induced dipole makes an angle of 45° with the field, corresponding to the field leading the polarisation by 45° at the frequency of maximum loss in a dielectric dispersion. The frequency range of the membrane-charging dispersion is 1 kHz to 1 MHz (increasing with medium conductivity), with frequencies up to 10 MHz expected if work in physiological saline were not impractical (due to thermal effects). Therefore this appears to be the single cell manifestation of the β dispersion seen when making dielectric measurements of bulk tissue (26).

It can be expected that rotation maxima will also be seen at other frequencies reflecting the dielectric properties of the material of the cell itself, or of collective properties (e.g. counterion movements) of the cell and medium. Evidence for the latter is that another rotation maximum for yeasts and bacteria occurs at approximately 100 Hz (as well as the rotation at higher frequencies which may correspond to counterion relaxation in the medium (27, 28, 29) or in the cell walls (30) of these organisms. However, the mechanism cannot be precisely analogous to that operating in the 1-1000 kHz range, because in that case the rotation is observed to be counter to the direction of the field (4, 16, 17) whereas the rotation of walled cells at low frequencies is seen to be co-field (in the same direction as the field). Co-field rotation of all cells so far examined by rotating field is also seen to have a maximum at approximately 1 MHz or higher. (Higher frequencies of rotating field are not yet available with sufficient precision).

Molecular dipole dispersions seem to be responsible for the multi-cell rotation of plant protoplasts (and of erythrocytes) which has been observed between 30 and 200 MHz (Küppers and Wendt, unpublished data). It is likely that the contribution of dipolar side chains can also be observed, as shown by the following. Chelating ion exchange ("Chelex") beads containing doubly-liganded doubly-negatively charged groups (tightly complexed with Ca^{++} or Cu^{++} ions) show multi-cell rotation in the 20-100 MHz range. This rotation is abolished, or shifted to much higher frequencies, when the sodium form (which is not expected to be a chelate) is examined. Frequencies seen to give strong positive dielectrophoresis produce no rotation, and it seems that in this case at least negative polarisability is necessary to allow rotation.

It can be expected that other dispersions, e.g. that of lipid head groups in membranes (31) and of water bound to proteins (32) and other materials should be observable. If this is so, then analysis of the differences in dielectric properties between individual cells should become a very powerful technique.

Conclusion

Extensive studies carried out on cell rotation have not confirmed the hypothesis (9, 18, 19) that the living cell has an intrinsic electric oscillator, at least between 100 Hz and some 200 MHz. We believe that this phenomenon is purely passive and not an example of a coherent excitation. However, it is not unusual in research for investigations designed to elucidate one idea to lead to the development of entirely different ones. In this case, we believe that cell rotation has already shown its worth as a research tool, and further developments are only a matter of time.

Acknowledgements

We are very grateful to our co-workers for provision of results prior to publication and to Mrs. S. Alexowsky for typing the manuscript. This work was supported by a grant of the BMFT (No. 03 7266) to U.Z.

References

1) Zimmermann, U., Scheurich, P., Pilwat, G. and Benz, R. (1981). Angew. Chem. 93, 332-351, Int. Ed. 20, 325-344
2) Zimmermann, U. and Vienken, J. (1982). J. Membrane Biol. 67, 165-182
3) Zimmermann, U. (1982). Biochim. Biophys. Acta 694, 227-277
4) Arnold, W.M. and Zimmermann, U. (1983). in Biological Membranes, Vol. V. Chapman, D., Ed. Academic Press, London, in press
5) Teixeira-Pinto, A.A., Nejelski, L.L., Cutler, J.L. and Heller, J.H. (1960). Exp. Cell Res. 20, 548-564
6) Füredi, A.A. and Valentine, R.C. (1962). Biochim. Biophys. Acta 56, 33-42
7) Füredi, A.A. and Ohad, J. (1964). Biochim. Biophys. Acta 79, 1-8
8) Pohl, H.A. and Crane, J.S. (1971). Biophys. J. 11, 711-727
9 Pohl, H.A. and Braden, T. (1982). J. Biol. Phys. 10, 17-30
10) Pohl, H.A. (1978). "Dielectrophoresis". Cambridge University Press, Cambridge
11) Mischel, M. and Lamprecht, I. (1980). Z. Naturforsch. 35c, 111-1113
12) Zimmermann, U., Vienken, J. and Pilwat, G. (1981). Z. Naturforsch. 36c, 173-177
13) Hub, H.-H., Ringsdorf, H. and Zimmermann, U. (1982). Angew. Chem. 94, 151-152; Int. Ed. Engl. 21, 134-135
14) Holzapfel, Chr., Vienken, J. and Zimmermann, U. (1982). J. Membrane Biol. 67, 13-26
15) Arnold, W.M. and Zimmermann, U. (1982). Naturwissenschaften 69, 297
16) Arnold, W.M. and Zimmermann, U. (1982). Z. Naturforsch. 37c, 908-915
17) Pilwat, G. and Zimmermann, U. (1983). Bioelectrochem. Bioenerg., in press
18) Pohl, H.A. (1980). in Bioelectrochemistry, Keyzer, H. and Gutmann, F. Eds., Plenum Press, New York
19) Pohl, H.A., Braden, T, Robinson, S., Piclardi, J. and Pohl, D.G. (1981). J. Biol. Phys. 9, 133-154
20) Noyes, R.M. (1980). Ber. Bunsenges. Phys. Chem. 84, 295-303
21) Lafon, E.E. and Pohl, H.A. (1981). J. Biol. Phys. 9, 209-217
22) Schwan, H.P. (1957). in Advances in Biological and Medical Physics, Vol. 5, J.H. Laurence and C.A. Tobias, Eds., Academic Press, New York, pp. 147-209
23) Jeltsch, E. and Zimmermann, U. (1979). Bioelectrochem. Bioenerg. 6, 349-384
24) Almers, W. (1978). Rev. Physiol. Biochem. Pharmacol. 82, 96-190
25) Takashima, S. and Schwan, H.P. (1965). J. Phys. Chem. 69, 4176-4182
26) Schwan, H.P. (1977). Ann. N.Y. Acad. Sci. 103, 198-213
27) Schwan, H.P., Schwarz, G., Maczuk, J., and Pauly, H. (1962). J. Phys. Chem. 66, 2626-2635
28) Schwarz, G. (1962). J. Phys. Chem. 66, 2636-2642
29) Dukhin, S.S. (1971). Surf. Coll. Sci. 3, 83-165
30) Einolf, C.W., Carstensen, E.L. (1973). Biophys. J. 13, 8-13
31) Shepherd, J.C.W. and Büldt, G. (1979). Biochim. Biophys. Acta 558, 41-47
32) Pennock, B.E. and Schwan, H.P. (1969). J. Phys. Chem. 73, 2600-2610

Symposium on Coherent Excitations in Biological Systems: Some Impressions and Conclusions

H. P. SCHWAN

Department of Bioengineering, University of Pennsylvania/D3, Philadelphia, PA 19104, USA

Some 40 scientists with background in the physical and biological sciences met from November 29 to December 1 in Bad Neuenahr, West Germany in order to discuss coherent excitations in biological systems and related topics. Field induced force effects, known as pondermotoric effects, and their biological manifestations were also considered. In addition several papers were offered on such topics as biological structure, metastable states in proteins, intracellular water, mm and sub-mm wave spectroscopy. Most of these papers are included in this volume. I shall therefore not attempt to summarize them, but rather limit myself to comment only on a few and draw some general conclusions. The presentations were limited to twelve papers. This and the effective chairmanship of Froehlich assured room for valuable disucssions and interactions.

Evidence for highly frequency specific biological effects of mm-waves had been presented several times during the past decade. The Russian reports from the Puchinov Biophysics institute of the USSR Adademy of Sciences were never published in detail and Webb's results had been debated on technical grounds. First detailed reports of highly frequency dependant growth of yeast cells were then published by Grundler and Keilmann. But their mm-wave applicator was complex and defied analysis of the specific absorption rate (SAR) in the test solution. Grundler has since then introduced a superior radiator. He presented his new results. There were three highlights in his presentation. First, very detailed data with more points then ever before demonstrated highly frequency dependant growth effects. Second, the new results confirm the results with the earlier radiating device. Third, observations on single cells and their response to the field also indicate high frequency specificity. The extension of his new technique of single cell observation appears promising. Motzkin from the Polytechnical Institute of New York briefly summarized her results, using different techniques and materials. She has so far not been able to observe frequency specificity. Nor has the Utah group headed by O. Gandhi which published their results in detail about a year ago. But Gandhi et al obtained biological responses at 0.5 GHZ intervals, while Grundler's results require resolution better than 10 MHZ. Others

Coherent Excitations in Biological Systems
Ed. by H. Fröhlich and F. Kremer

should try to duplicate Grundler's results, using the same technique and biological endpoint. Until this is done, some will continue to debate the significance of Grundler's results. If high frequency specificity will be accepted beyond doubt, Froehlich's theory of coherent excitation in the mm-wave region will probably account for such results. The implications to biophysical research are then no doubt enormous even though at the moment unpredictable.

Some other papers in this category of mm-wave effects were presented by Kremer and Genzel. Kremer observed non-thermal effects of mm-wave radiation on giant chromosomes. Since his frequency was swept between 64 and 69 GHz, high frequency specificity is not necessarily indicated. Genzel and coworkers studied the dynamics of large biomolecules from 50 to 150 Ghz and in the range of 20 to 480 wavenumbers per cm over an extended temperature range. Strong absorption peaks were observed and a relaxation model proposed. But, as Genzel stated, these phenomena will probably not take place if biomolecules in an aqueous medium are used because of the strong absorption of liquid water.

Hasted also presented absorption data in the 10 to 550 cm^{-1} range. Proteins, polysaccharides and amino acids were investigated. The data demonstrated that absorption features of the more complex proteins can be composed at least in part from those of their subunits. Furthermore it appears that hydrogen bond vibration fundamentals and their harmonics are involved.

Hasted presented evidence for the existence of metastable states in proteins. The existence of such metastable states have been suggested by Froehlich. Hasted demonstrated that the dielectric properties of hemoglobin films consisting of 3 to 7 molecular layers display relaxation effects below some Hertz which are strongly dependant on the applied field strength of some MV/m. Such field strength values exist in biological membranes. Properties acquired under the influence of the applied field return slowly over weeks to the original level. This writer mentioned at the meeting that electrode polarization processes might participate in the low frequency relaxation process. But it appears on closer examination of the data provided by Dr. Hasted that this is less likely than originally anticipated. Clegg presented a detailed summary of cytoplasmic organelles and the cytoskeleton discovered only in recent years. He estimated the surface area of all such structures in a cell. Assuming structured water to extend up to 100 or 150° from the surface he concluded that most water inside cells ought to be different from normal water. However if one assumes bound water to be restricted to a few molecular layers the usual amounts of bound water would probably be arrived at.

The last half day of the meeting was largely devoted to field induced forces and their biological manifestations. These pondermedoric effects can cause many biological responses, are of a nonthermal nature and also frequency dependant, even though this dependence is not as sharp as anticipated by Froehlich theory of mm-wave interactions with biosystems. Pohl, who studies extensively the movement of particles and cells in non-uniform alternating electrical fields gave an excellent summary of it. Zimmermann, known for his past work on the breakdown of membranes under electrical stress, has concerned himself in more recent years with a variety of cellular responses to alternating and pulsed fields. He gave a stimulating presentation and demonstrated how useful the study of these effects is for a variety of purposes. Sauer presented a paper summarizing his derivation for the force acting on a particle exposed to an alternating electrical field and the trajectory of a particle approaching another one. The force equation was derived for the general case that both particle and medium have complex permittivity, thereby extending previous equations for the pure dielectric case. This finally should put to rest a longstanding debate what equations to use in the complex case.

Biological interaction mechanism of electromagnetic radiation whose quantum energy is far too low to form radicals have been of increasing interest during the past two decades. Froehlich proposed more than a decade ago that biological reactions to nonionizing radiation are likely to be nonlinear and highly frequency specific. Indeed if such interactions were linear then the strong absorbance of water would prevent a resonant response. So far it has not been possible to demonstrate specific sites at the cellular or macromolecular level which could be responsive. Suggestions include biological membranes, biological macromolecules, the intracellular cytoskeleton, structured water near macromolecular surfaces. The papers in this volume contribute to the debate even though they can not pinpoint yet a specific interaction site . They contain valuable material likely to influence future research.

The meeting was well organized and supported financially by IBM Deutschland GmbH Stuttgart. It took place in a pleasant well known West German spa and facilities were excellent. Thanks are due to IBM and Professor Froehlich for making this meeting possible.

MIX
Papier aus verantwortungsvollen Quellen
Paper from responsible sources
FSC® C105338

If you have any concerns about our products,
you can contact us on
ProductSafety@springernature.com

In case Publisher is established outside the EU,
the EU authorized representative is:
Springer Nature Customer Service Center GmbH
Europaplatz 3, 69115 Heidelberg, Germany

Printed by Libri Plureos GmbH
in Hamburg, Germany